THE METAPHYSICS OF SCIENCE

CRAIG DILWORTH
Department of Philosophy, Uppsala University

THE METAPHYSICS OF SCIENCE

*An Account of Modern Science
in Terms of Principles, Laws and Theories*

SECOND EDITION

 Springer

A C.I.P. Catalogue record for this book is available from the Library of Congress.

ISBN 978-1-4020-6327-5 (PB)
ISBN 978-1-4020-3837-2 (HB)
ISBN 978-1-4020-3838-9 (e-book)

Published by Springer,
P.O. Box 17, 3300 AA Dordrecht, The Netherlands.

www.springer.com

Printed on acid-free paper

First edition 1996

"Perhaps the greatest theoretical success of the book is its convincing arguments for the importance of metaphysical considerations to science. On the PTL model, Dilworth shows why metaphysics plays, or should play, a central role in understanding modern science. If one accepts the notion that there are non-empirically testable (in the standard sense) principles on which all of science is based, one is also forced to admit metaphysics. ... Dilworth does a convincing job of walking the fine line between the strong foundationalist and strong coherentist positions, carving out, in the process, an interesting place for metaphysics in science."

Dialogue

"The book would make a suitable addition to the shelves of any philosophy department or science faculty library. It is thoughtful, original and many will find it provocative."

Australasian Journal of Philosophy

"Dilworth does not content himself with a mere philosophical analysis of the phenomenon of modern science, but tries to draw a lesson from this analysis applicable to the actual practice of science. Whereas in its beginnings modern science was a paradigm of open-mindedness, it is now in danger of becoming an ideology, due to its refusal to reflect on its own principles. *The Metaphysics of Science* performs the much-needed function of opening the door to such reflection – both for professional philosophers and scientist themselves."

Epistemologia

"The book is clearly written and well structured. [It provides] an interesting general introduction to the philosophy of science from quite a different perspective than is usually offered – a perspective which is decidedly Kantian in flavour."

Philosophy

"The main message of *The Metaphysics of Science* is that science is based on metaphysical principles. [Dilworth] expresses this in an apposite way such that scientific concepts are 'principle-laden,' where the principles intended are those of the uniformity of nature, the continuous existence of substance, and causality. Around this he builds a model of scientific explanation, based on the three concepts, principles-theories-laws, and shows its applicability in a number of areas of science."

Sven Ove Hansson, University of Stockholm

"Dilworth's book is [an] interesting Whewellian 'top-down' account of the aims, methods and structure of science. An outgrowth of *Scientific Progress*, this monograph is a very well written and insightful account of science as an enterprise which imposes certain (intelligible) principles on more specific theories."

Matti Sintonen, University of Tampere

"These two works [*Scientific Progress* and *The Metaphysics of Science*], the latter growing out of the former, are a sustained and comprehensive account of the methodology of science in the light of the theories which have dominated twentieth-century thinking. Dilworth argues that all the competing accounts are flawed, and substitutes his own, applying it in detail to concrete examples from both the natural and social sciences. ...

"Both [books] have much to recommend them. They are written with great fluency, and their combined historical survey is immensely valuable. They also contain a wealth of critical comment which ... is important and suggestive, providing a stimulus for further debate. Together they make a significant contribution to the philosophy of science, and will be found useful to both students and professionals alike."

Ratio

CONTENTS

PREFACE TO THE FIRST EDITION

The roots of this work lie in my earlier book, *Scientific Progress*, which first appeared in 1981. One of its topics, the distinction between scientific laws and theories, is there treated with reference to the same distinction as drawn by N. R. Campbell in his *Physics: The Elements*. Shortly after completing *Scientific Progress*, I read Rom Harré's *The Principles of Scientific Thinking*, in which the concept of theory is even more clearly delineated than in Campbell, being directly connected to the notion of a *model* – as it was in my book. In subsequent considerations regarding science, Harré's work thus became my main source of inspiration with regard to theories, while Campbell's remained my main source with respect to empirical laws.

Around the same time I also read William Whewell's *Philosophy of the Inductive Sciences*. In this work, Whewell depicts *principles* as playing a central role in the formation of science, and conceives of them in much the same way as Kant conceives of fundamental synthetic a priori judgements. The idea that science should have principles as a basic element immediately made sense to me, and from that time I have thought of science in terms of laws, theories and principles.

Two questions then presented themselves, namely precisely how laws and theories are related to principles, and what the fundamental or core principles of modern science actually are. Though an answer to these questions had already been put forward by Whewell, his conception was in a particular respect too rigid, and furthermore lacked a certain clarity and simplicity. It has been mainly through attempts to overcome these shortcomings that the present work has taken form. Here, modern science is conceived as an enterprise centred on three particular principles in such a way that it includes as integral aspects both empirical laws and abstract theories.

After already having become clear as to the basic structure of this work, I read F. S. C. Northrop's *Science and First Principles*, in which Greek scientific-philosophical thought is conceived as consisting of three main streams, with modern science being a development of one of them in particular. Northrop's historical view not only fit well with my philosophical conception, but was also in keeping with my own thoughts on these matters.

So the present work has been influenced mainly by Whewell, Campbell and Harré, and from one point of view may be seen as a synthesis of certain of their central ideas in such a way as is in keeping with the historical account of Northrop; from another point of view it may be seen as a development of Chapter 10 and Appendices I and IV of the latest edition of *Scientific Progress*. It is a work, like *Scientific Progress*, fundamentally antithetical to the logico-linguistic tradition of twentieth-century analytic philosophy, and moreover one which constitutes an instance of the application of a metaphysical approach in the philosophy of science.

A number of people have helped me with comments on earlier publications and on the present text at various of its stages, and to all of them I extend my thanks. They include Evandro Agazzi, John Blackmore, Robert S. Cohen, Yves Gingras, Rom Harré and Peter Manicas. Peter Söderbaum and Stig Wandén have commented on Chapter 6, as has Roger Pyddoke, with whom I have previously worked on the topic of that chapter. Comments on Chapter 7 have been afforded by Paul Dumouchel and Jaap van Brakel. I would also like to thank James Crompton, Ingvar Johansson and Giovanni Sommaruga for comments on the whole of the book in manuscript. Very special thanks are due to Louk Fleischhacker, who has discussed the subject of the book with me in detail, and has been a source of encouragement throughout its preparation. Financial support for research has been provided by the Swedish Council for Research in the Humanities and Social Sciences.

STOCKHOLM
May 1995   C. D.

PREFACE TO THE SECOND EDITION

Apart from typographical improvements, the addition of the odd footnote, and the emendation of a small part of the text, this edition has been supplemented with three appendices. The first of these provides justification for claims made at the end of Chapter 10 regarding the future of humankind. In this appendix, a theory of human development is presented based on what is termed *the vicious circle principle*. According to the theory, the key difference between the development of humans and other animals lies in our ability to generate new technology. This ability has meant that on many occasions we have been able to solve problems of need by technological means, which has led to a growth in the human population, which in turn has brought us back to a situation of need, which we often once again succeed in overcoming through technological innovation. The vicious aspect of the circle consists in its leading to a dead end when either resources no longer exist that are amenable to technological innovation, or the waste the ever-increasing use of technology gives rise to makes further development impossible.

The second appendix is an analysis of particular enterprises which we intuitively take to be non-scientific against the background of the principles of modern science as they are presented in this book. This analysis should perform either or both of two tasks, namely that of indicating the extent to which the various subjects treated are or are not scientific, and/or that of supporting the approach of the present work by showing how its handling of this question is in keeping with our intuitions.

The third appendix is a comprehensive and detailed reply to criticisms made of this book and of later editions of my other major work in the philosophy of science, *Scientific Progress*. In it I show that my critics have been unable to move beyond an essentially logical-empiricist conception of science – to which alternative conceptions are provided in both books – which has prevented them from properly appreciating the views being advanced in either of them.

I would like to thank Jan Faye for comments on Appendix II, and Louk Fleishhacker for comments on all three appendices.

STOCKHOLM
August 2005   C. D.

INTRODUCTION

This book, in its attempt to depict the metaphysics of science, has a form which differs in a number of ways from that of most other contributions to the philosophy of science. Part of this difference is already implied in the book's title, for few modern writers would want to say that science has any metaphysics at all. What does it mean to say that science has a metaphysics?

Metaphysics itself may be thought of as having two main aspects which, following Kant, we shall call the transcen*dent* and the tran-scenden*tal*. Both of these notions are important for the message of the present work. The transcendental, as it is to be understood here, may be seen as consisting of a person's most fundamental convictions or beliefs about the nature of reality. These are beliefs – such as, for example, belief in the existence of God – which affect the whole of a person's conception of reality, and which, psychologically speak-ing, are the most difficult to give up. Furthermore, they are beliefs of which a person may not be conscious. In distinction from Kant's view, however, we do not take the transcendental to be independent of experience; rather, while the beliefs that compose it are not just generalisations of experience, they are nevertheless arrived at through some combination of experience and thought. What is important how-ever is that once arrived at, they constitute the very preconditions for the way one afterwards experiences the world. And, as follows from this, the transcendental need not take a predetermined form, as it did for Kant. As conceived by him, the very constitution of humans was such that they experience nature in terms of, for example, cause and effect. Here, the transcendental is open to reform – in terms of our example, though one's belief in God may be deep, it may come to be given up.

As regards science then, the transcendental may be seen as con-cerning the most fundamental beliefs scientists as a group have re-garding the nature of reality, as these beliefs are manifest in their scientific endeavours. Or, moving from psychology to epistemology, we should say that the transcendental for science consists of the most fundamental *presuppositions* of science. In being transcendental with respect to science they cannot have been arrived at through the pursuit of science, but must be, in a definite sense, pre-scientific, or metascientific. And, as in the example given above, they can be revised or abandoned in favour of alternatives. Thus the idea that science has transcendental presuppositions does not conflict with the idea that science is a dynamic, evolving enterprise, but rather directs the philosopher's attention to analysing the dynamics and evolution of science in terms of changes in its presuppositions. These changes may be more or less drastic, which can lead us to say that changes in science, or scientific revolutions, are more or less total. And at the end of the day we may still find that certain basic presuppositions – or core *principles* – have continued to function as regulative ideas for science throughout its history, and furthermore that this transcendental heart of science is what makes science what it is and not another thing. It is just this transcendental aspect of science, and how it affects the enterprise of science as a whole, that is focussed on in the present work.

The other aspect of metaphysics is the transcendent, which, broadly speaking, is that which lies beyond the limit of some generally accessible realm, whether it be, for example, experience, knowledge, understanding, language, or thought. Of immediate relevance for the study of science is the idea of something's lying beyond the limit of knowledge, where knowledge is understood as empirical knowledge. It is largely with regard to the issue of whether science should be restricted in its investigations to what can be empirically known, or whether it should also delve into the transcendent realm of theoretical entities, that the long-standing battle in the philosophy of science between empiricists and realists has been waged (whereas the issue concerning the transcenden*tal* may be seen as the focal point of the debate between empiricists and rationalists). Our difference from Kant with regard to the transcendent, apart from his not considering its possible application as limiting realms other than that of empirical knowledge, is that what is transcendent at one point in time need not remain so – what is at one time hypothetical may become factual. So we have more of a pragmatic view of the transcendent than does Kant, at least as regards its application to science. And further, on our notion there are *levels* of transcendence, such that, for example, physical atoms may be viewed as transcendent with respect to particular empirical laws concerning gases, while quarks and leptons may be considered transcendent with respect to physical atoms. It may here be noted that, properly understood, the realist is not necessarily advocating that one can have knowledge of any particular transcendent realm, but that it is through theorising about the nature of such a realm and its relation to the non-transcendent or empirical that the latter can be made intelligible. This issue is the topic of the first chapter of the book, and constitutes a theme throughout the work which is rounded off in the penultimate chapter.

So what is here meant by saying that science has a metaphysics is that it has a transcendental aspect, the question of whether it also has or ought to have a transcendent aspect being one investigated in the book against the background of the presupposition that the transcendental aspect has the particular form specified in Chapter 2.

A second way in which this book differs from most other contributions to the philosophy of science is in its emphasis on 'paradigm-thinking.' There are various ways in which the relation between the transcendental and science may be conceived, and the way in which it is conceived here is perhaps novel. During the nineteenth century it was common among philosophers and scientists to think in terms of the principles of science, but for them the principles were to constitute the *basis* of science, whereas here principles are to constitute the *core* of science – a distinction to which we shall return directly. Furthermore, for most thinkers at that time, and even today, science was conceived of as a monolithic enterprise providing the one and only sure route to truth. Here, on the other hand, as is in keeping with the view that the transcendental can take different forms, science is conceived as one particular epistemological activity which may be compared with others. Moreover, what is intended by *science* in the present work is restricted to what is normally considered *modern* science, i.e. science since the time of Galileo and the Scientific Revolution. Thus modern science can be compared with other ostensible means of gaining knowledge or understanding of reality – such as Aristotelian science, or magic – such a comparison to be made first in terms of similarities and differences in their core principles.

What is meant by saying that particular principles constitute the core rather than the basis of science is that they are not general self-evident truths from which particular empirical truths can be formally deduced, but are rather ideal conceptions of reality which guide scientists' investigation of actual reality. From this perspective, what makes a particular activity scientific is not that the reality it uncovers meets the ideal, but that its deviation from the ideal is always something to be accounted for. In this way transcendental principles constitute *paradigms* in much the same sense as this term is intended by Kuhn (and Wittgenstein), it being understood however that they are conceptual and ontological rather than concrete and methodological in nature. Thus, as distinct from Kuhn's view, principles constitute a paradigm for modern science in that they are mental constructs depicting, in broad outline, an ideal reality, rather than being instances of scientific practice embodying an ideal method. Similarly to Kuhn's view, on the other hand, an enterprise focussed on a particular ontological paradigm can go through a number of historical phases. On the present account, the paradigm of modern science as a whole had its golden period during the nineteenth century – a period that has been termed 'the age of science' – while during the twentieth century greater difficulty has been experienced in the attempt to assimilate the results of scientific enquiry to its transcendental ideal, particularly in its core discipline of physics.

Speaking of the 'core discipline of physics' brings us to another sense in which transcendental principles constitute the ontological paradigm of modern science. Thus, just as science had an historical period during which the reality it revealed was most similar to the ideal depicted in the principles, so too can we say that various scientific disciplines lie closer or further from the transcendental core of science depending on how similar the reality they uncover is to the reality of the principles. Due to the nature of the principles of science on the one hand, and reality on the other, the greatest success in applying the principles has been had in physics and chemistry, while biology lies further from the core, and the social sciences further still.

Another way that paradigm-thinking enters the present work is in the claim that this sort of thinking actually occurs in science. Thus on the present view both scientific theories and the expressions of empirical laws constitute in science itself intellectual paradigms intended to capture the *essence* of particular aspects of reality; and these essences are not necessary or sufficient conditions that reality must meet in order for the relevant laws or theories to be applicable, but *idealised* states of affairs which as a matter of fact might never have real correlates. One area in which this paradigm-thinking is particularly clearly manifest is in the treatment of natural kinds in biology (as examined in Chapter 7), where difference of natural kind is not an all-or-nothing affair. In this context, paradigm-thinking involves the taking of certain real things or intellectual constructs as each constituting the ideal of a particular type, such that individual entities are seen as gravitating more or less to one paradigm or another depending on their characteristics; in this way such an entity may thus be considered to be of some particular type, or perhaps to constitute a borderline case between types.

The use of paradigm-thinking in the present work does not stop there however, but lies in the background throughout. Thus, when in the book we speak of the function of theories as being to provide causal explanations of laws, we mean that this is their *paradigmatic* function, which does not exclude their being used, for example, to provide information about a deeper-lying reality as such; or when we say that the aim of the empirical aspect of science is the discovery of empirical laws, we mean this is its *paradigmatic* aim, and do not intend to deny that the empirical aspect of science may also involve e.g. the determination of the existence of particular entities. The conception of science presented here is, it is hoped, a coherent whole in which various concepts occupy particular nodal points, thereby making it also a system. In this system these concepts, the most important of which are principles, laws and theories, function at these nodes as conceptual paradigms.

So the notions of principles, laws and theories constitute the nodal or paradigm concepts in terms of which the present account of science is conducted. This account, in broad outline, runs as follows. Modern science, as presented in Chapter 2, is a particular epistemological enterprise which consists in the application of, and thereby obtains its nature from, particular *fundamental metaphysical principles*. In order to find clear application to reality, these fundamental principles are refined in various ways, giving rise to different groupings of what may be termed *refined principles*, each grouping defining a different science or scientific discipline. Where the fundamental principles are normally implicit in the doing of science, the refined principles are explicit.

The assumption that reality has the basic nature depicted by the implicit fundamental principles leads to its being investigated according to a particular method – the experimental method – resulting in the discovery of *empirical laws* (the topic of Chapter 3). There is no guarantee however that the laws discovered by employing this method will be in keeping with the fundamental principles as a whole, nor, more particularly, with the refined principles of the science or scientific discipline in question. In order to show that and how they are so, one or more depictions of the reality being investigated is advanced, depictions each of which is more detailed than that provided by the refined principles.

Such depictions are of hypothetical realities which on the one hand naturally give rise to the empirical laws that are of interest, while at the same time are constrained by limits set by the refined principles. This constraint consists in the depicted realities' not transgressing the refined principles, as well as in the depictions' employing only concepts taken from them. Such ontological depictions are *theories* which, if they achieve their aim, may be said to have scientifically *explained* the laws in question by showing them to be but an empirical manifestation of the principles underlying the science in question (the topic of Chapter 4). In this way, where empirical laws provide scientific *knowledge*, theories, by linking the laws to the principles, provide scientific *understanding*.

This, then, is the central message of this book. Following its presentation in Chapters 2 to 4, it is employed in various ways. First, in Chapter 5, it provides the structure for a model of scientific explanation. This model is the Principle-Theory-Law (PTL) model, which involves a further development of the law/theory distinction in introducing notions of the *nominal* vs. the *real* aspects of the domain of a theory. While it is intended that this model capture the essence of explanation in modern science, it is possible that it also has application outside this realm. In any case, it is applied in Chapter 6 to a case study taken from modern microeconomics, where it appears to fit rather well. There, according to the core ideas of the book, the key difference between mainstream economics and the natural sciences, apart from their subject-matter, lies in their having different conceptions of causality.

Against the background of one of the fundamental principles of science presented in Chapter 2, the distinction between the nominal and the real developed in the PTL model is presented in Chapter 7 as the key to understanding the modern-scientific conception of natural kinds. There it is suggested that for modern science natural kinds are to be conceived of as having both nominal and real essences, where a real essence can be a nominal essence relative to some even deeper-lying real essence. This notion of difference of level, mentioned above with regard to the transcendent, stems from the relativisation of the law/theory distinction; it is also a key aspect of the discussion of probability in Chapter 8. In that discussion, for which another fundamental principle is central, the distinction is made between nominal and real probability determinations, where nominal probability is based on empirical samples while real probability is based on ontological theories.

In Chapter 9, as mentioned, the realist/empiricist discussion is brought to a close; and the distinctions between epistemology and ontology, and between knowledge and understanding, are discussed. In the final chapter, Chapter 10, the modern-scientific worldview is compared with other historical worldviews, and the question is raised whether for both epistemological and pragmatic reasons it may be time to change to some fundamentally different epistemology, whether or not the name "science" be applied to it.

As is clear from the above, this work is one in the philosophy of science as distinct from epistemology. We are here focusing on the nature of modern science, and not on how best to obtain knowledge of reality in some wider context. Though basic epistemological issues are broached in Chapter 3, this is done only with the aim of determining the fundamental conception of knowledge acquisition of modern science. This marks a third difference between this work and most other contributions to the philosophy of science. In them it is implicitly assumed that modern science constitutes the best way of obtaining knowledge about reality, the question being what is the best way to conduct science. In other words there is an implicit faith that humankind has been constantly moving forward along the one

road to Truth, that road being Science, without consideration being given to the thought that modern science might appear just as
wrongheaded in the future as alternative forms of science do now.
The interest of the present volume, on the other hand, is primarily in
clarifying the nature of science, though questions as to its value are
taken up in the final chapter.

As is implied in that chapter, that the nature of modern science be
clarified is becoming a pressing need in an age where science, while
helping provide most people in industrialised nations with a high
standard of living, has at the same time been an essential factor in the
development of nuclear and chemical weapons, and a contributing
factor to the spread of pollutants threatening future human life on
earth. In spite of these trends, science shows signs of becoming the
first world-wide religion. The scientific enterprise is in serious need
of demystification.

The present work is intended to reveal the nature of modern science
as an intellectual enterprise. If successful, it should thereby explain
such central aspects of modern science as the way in which the Scientific Revolution of the seventeenth century actually was a revolution
with respect to preceding epistemological approaches. Similarly, it
should explain both why the nineteenth century may be considered
the golden age of science, as well as why physics and chemistry lie at
the core of the enterprise while the social sciences continue to struggle
to obtain scientific status. It should also afford a means of demarcating science from non-science (in terms of paradigms), and explain
the nature of scientific revolutions, whether major or minor. Furthermore, the present work should clarify the nature of the foundational
problems in physics today, as well as the nature of such activities as
scientific explanation and classification.

We begin our excursion into the metaphysics of science by examining the historical debate regarding the role of the transcendent in
science, as manifest in the empiricism/realism issue in the philosophy
of science.

CHAPTER 1

EMPIRICISM VS. REALISM –
THE PERENNIAL DEBATE
IN THE PHILOSOPHY OF SCIENCE

The issue over empiricism and realism, presently the focus of much discussion in the philosophy of science, is but the manifestation of an age-old perplexity. The perplexity is over the relation between one's experiences, and the world, or between phenomena and reality.

Now of course phenomena may themselves be considered real, and the realm they constitute be considered a world. But such a world differs from the reality with which it is being contrasted here by the fact that the latter is conceived to be the only one of its kind, to exist independently of being experienced, and to be common to us all.

At one time or another virtually every conceivable line has been taken on the issue, from the view that there is no reality other than phenomena, to the view that reality, while different from phenomena, alone causes and is perfectly represented by them. The interest of the present chapter will be in presenting some of the more conspicuous forms the issue has taken in the philosophy of science since the time the subject began as a relatively autonomous discipline, and in critically appraising some of the more recent contributions against the background of the debate as a whole.

Empiricism in the philosophy of science is, broadly speaking, the view that scientific investigation be confined to phenomena and their formal relations, while realism is the view that it is also the task of science to investigate the *causes* of such phenomena and relations, conceived as emanating from the real world.[1] **Auguste Comte**

---

[1] This is the conception of realism involved in the historical debate. As will be argued below, the modern conception of 'scientific realism,' according to which theories are statements and the aim of science is to produce theories ever closer to the truth, is essentially an empiricist conception, and fails to come to grips with the

(1798–1857) strongly advocates a variant of the empiricist view, which, following Saint-Simon, he terms *positivism*:

Our study of nature is restricted to the analysis of phenomena in order to discover their *laws*, that is, their constant relations of succession or similitude, and can have nothing to do with their *nature*, or their *cause*, first or final, or the mode of their production. Every hypothesis that strays beyond the domain of the positive can merely occasion interminable discussions, by pretending to pronounce on questions which our understandings are incompetent to decide.[2]

Comte's main point is epistemological, not ontological, which we shall see also to be the case with most commentators on the issue in the philosophy of science. He is not denying that the domain of phenomena has a causal basis in a world transcending such phenomena (and might thus be considered an ontological realist by implication), but is arguing that the inaccessibility of this world makes it idle to posit hypotheses concerning its nature.

Comte sees his positivistic approach to the investigation of nature as the third stage of a process which first passes through theology and then metaphysics. In metaphysics we find a "vain search after absolute notions, the origin and destination of the universe, and the causes of phenomena."[3] He asks:

What scientific use can there be in fantastic notions about fluids and imaginary ethers, which are to account for phenomena of heat, light, electricity and magnetism? Such a mixture of facts and dreams can only vitiate the essential ideas of physics, cause endless controversy, and involve the science itself in the disgust that the wise must feel at such proceedings. ... These hypotheses *explain* nothing. For instance, the expansion of bodies by heat is not *explained* – that is, cleared up – by the notion of an imaginary

---

epistemological issue. In this regard, cf. Sober (1980), p. 371, and Blackmore (1983), p. 34.

[2] Comte (1830–42), p. 147.

[3] Ibid., p. 72. Next quote, pp. 148–149. In spite of this view of metaphysics, however, Comte claims elsewhere that "the human mind could never combine or even collect [*recueiller*] observations unless it were directed by some previously adopted speculative doctrine" (*Discours sur l'Esprit Positif*, Paris, 1844, p. 6; cited in Laudan, 1981, p. 146). Another point difficult to reconcile with the main thrust of his work, namely an instrumentalistic acceptance of theorising, is taken up in Laudan (1981), pp. 150–156.

fluid interposed between the molecules, which tends constantly to enlarge their intervals, for we still have to learn how this supposed fluid came by its spontaneous elasticity, which is, if anything, more unintelligible than the primitive fact.

For Comte, explanation is not to involve the positing of the existence of unobservable entities: "an explanation of facts is simply the establishment of a connection between single phenomena and some general facts, the number of which continually diminishes with the progress of science,"[4] though it is not evident whether Comte expects explanation in this sense to suffice to 'clear up' e.g. why heated bodies expand. Since such connections and the relations constituting the laws of phenomena have nothing to do with the cause of the phenomena, neither they nor the explanations in which they figure are causal. Here we see how the view that the only relations of epistemological relevance are formal is a natural consequence of empiricism (positivism), and serves to complement it. Comte, however, does not see these relations as being of formal logic, but of mathematics, "the most powerful instrument that the human mind can employ in the investigation of the laws of natural phenomena." On this view, then, such issues as that concerning action at a distance do not arise, for there is no requirement to indicate the cause of the phenomena in which this principle appears operative, but merely to show the phenomena to be formally connected to some general facts.

Comte's opposition to metaphysics and the investigation of causes is antithetical to the position of **William Whewell** (1794–1866). On Whewell's view, the notion of *force* in physics is nothing other than the empirical specification of the more general notion of *cause*. Thus, for example, each of Newton's three axioms or laws of motion is the empirical determination of a more fundamental causal maxim – a maxim whose empirical determination could take some other form in a different field of inquiry. Newton's first law of motion – that a body will continue in its state of rest or uniform motion in a straight line unless acted upon by a force – is in this way considered

---

[4] Comte (1830–42), p. 72; next quote, p. 100.

by Whewell to be an empirical specification of the maxim that nothing can take place without a cause.[5]

With reference to Comte, Whewell says:

Again, some persons condemn all that we have here spoken of as the discussion of ideas, terming it *metaphysical*: and in this spirit, one writer has spoken of the 'metaphysical period' of each science, as preceding the period of 'positive knowledge.' But as we have seen, that process which is here termed 'metaphysical' – the analysis of our conceptions and the exposure of their inconsistencies – (accompanied with the study of facts) – has always gone on most actively in the most prosperous periods of each science. There is, in Galileo, Kepler, Gassendi, and the other fathers of mechanical philosophy, as much of *metaphysics* as in their adversaries. The main difference is, that the metaphysics is of a better kind; it is more conformable to metaphysical truth.

As distinct from metaphysical truth, Whewell distinguishes two kinds of *inductive* truth: laws of phenomena, and theories of causes – the former describing an order which the phenomena follow, and the latter explaining *why* they follow that order by indicating the cause or causes of which the order is the effect.[6] Thus Kepler discovered that the planets describe ellipses, and Newton explained why they do so by reference to his (causal) laws of motion and the law of gravitation. But it may be that our empirical specification of causes be improved upon with the progress of science, so that what we take at any one time as the particular cause of a certain phenomenal order be mistaken. Again with reference to Comte, Whewell says:

Since it is thus difficult to know when we have seized upon the true cause of the phenomena in any department of science, it may appear to some persons that physical inquirers are imprudent and unphilosophical in undertaking this Research of Causes; and that it would be safer and wiser to confine ourselves to the investigation of the laws of phenomena, in which field the knowledge we obtain is definite and certain. Hence there have not been wanting those who have laid it down as a maxim that 'science must study only the laws of phenomena, and never the mode of production.' But it is

---

[5] Whewell (1847), Part 1, pp. 217–218; quote following, Part 2, p. 378.

[6] Whewell's distinction between metaphysical maxims (or axioms, or principles), laws of phenomena, and theories of causes is essentially the same distinction as that made in the present study between principles, laws and theories.

easy to see that such a maxim would confine the breadth and depth of scientific inquiries to a most scanty and miserable limit. Indeed, such a rule would defeat its own object; for the laws of phenomena, in many cases, cannot be even expressed or understood without some hypothesis respecting their mode of production. How could the phenomena of polarization have been conceived or reasoned upon, except by imagining a polar arrangement of particles, or transverse vibrations, or some equivalent hypothesis? . . .

To debar science from enquiries like these, on the ground that it is her business to inquire into facts, and not to speculate about causes, is a curious example of that barren caution which hopes for truth without daring to venture upon the quest of it. This temper would have stopped with Kepler's discoveries, and would have refused to go on with Newton to inquire into the mode in which the phenomena are produced. It would have stopped with Newton's optical facts, and would have refused to go into the mode in which these phenomena are produced. And, as we have abundantly shown, it would, on that very account, have failed in seeing what the phenomena really are.[7]

On the other hand there is a particular point on which it may appear that Whewell and Comte would agree, namely that regarding the fruitlessness of positing the existence of what we would today call theoretical entities. With regard to the atomistic theories prevalent at his time, Whewell says, "in explaining the properties of matter as we find them in nature, the assumption of solid, hard, indestructible particles is of no use or value." An attitude shared not only by Whewell and Comte, but by almost all early contributors to the debate, is a predilection for the attainment of *certainty*,[8] and this disposes Whewell to view the atomistic doctrine with a critical eye. But it should be noted that in the present context Whewell is not actually denying that something like atoms may exist and play a causal role in the manifestation of phenomena – his concern rather is over the claim that such entities be indestructible. Ironically, Comte himself appears to accept the notion of atoms where he suggests that "We say that the general phenomena of the universe are *explained* by [the doctrine of gravitation], because it connects under one head the whole immense variety of astronomical facts, exhibiting the con-

---

7 Ibid., Part 2, pp. 103, 104; next quote, Part 1, p. 438.
8 On the philosophical background to this predilection, see Manicas (1987), pp. 12–15.

stant tendency of atoms towards each other in direct proportion to their masses, and in inverse proportion to the squares of their distances;"[9] a doctrine which he considers to be a mere extension of the perfectly familiar fact of the weight of bodies on the surface of the earth.

J. S. Mill (1806–1873), while agreeing with Comte on fundamentals, sees a problem with his approach as lying in his objecting to the application of the word "causal" to relations of succession. In the spirit of Hume and in defiance of common sense, Mill at one point identifies causality with succession, pronouncing a Law of Causation, which "is coextensive with the entire field of successive phenomena, all instances whatever of succession being examples of it."[10] However, like Comte, he does not wholly deny the existence of causes which, in his terms, actually produce their effects, but considers them to be of no relevance to induction: "I premise then, that when in the course of this inquiry I speak of the cause of any phenomenon, I do not mean a cause which is not itself a phenomenon; I make no research into the ultimate or ontological cause of anything. ... The only notion of a cause which the theory of induction requires, is such a notion as can be gained from experience."

Thus for Mill, all laws of nature are causal in his sense, and no distinction is to be drawn on that ground between e.g. Kepler's laws and what is commonly termed Newton's theory of gravitation. Their indistinguishability on this ground, however, is not to exclude our speaking of Newton's 'theory' *explaining* Kepler's laws, for explanation on Mill's view consists in resolving laws into other laws. But explanation for Mill, unlike for Whewell, is not to answer the question *why?*

The word explanation is here used in its philosophical [*earlier eds.*: a somewhat peculiar] sense. What is called explaining one law of nature by another, is but substituting one mystery for another; and does nothing to render the general course of nature other than mysterious: we can no more assign a *why* for the more extensive laws than for the partial ones. The ex-

---

[9] Comte (1830–42), p. 75.
[10] Mill (1881), Bk. III, Ch. v, § 1; next quote, § 2. In this regard cf. McCloskey (1971), pp. 39–40: "Mill's concept is not the ordinary concept and ... the account he offered is not what is ordinarily meant by 'cause.'"

planation may substitute a mystery which has become familiar, and has grown to *seem* not mysterious, for one which is still strange. And this is the meaning of explanation, in common parlance.[11]

Another notion Mill treats in a noteworthy way is that of being physical. The notion of physical is normally applied to what is considered to be real and not merely phenomenal, but Mill uses it in terming the relations of succession between phenomena *physical* causes.[12]

Ernst Mach (1838–1916) also uses the term "physical" in a phenomenalistic sense, where he not only speaks of physical *experiences*, which are to consist solely of sensations, but goes so far as to identify physical objects (bodies) with sensations.[13] Mach, like Comte and Mill, is an empiricist; but his empiricism is more extreme than theirs, being ontological as well as epistemological. For him, the world consists solely of sensations and mathematical relations among them. "What we represent to ourselves behind the appearances exists *only* in our understanding, and has for us only the value of a *memoria technica* or formula, whose form, because it is arbitrary and irrelevant, varies very easily with the standpoint of our culture."[14] Thus, "In conformity with this view the ego can be so extended as ultimately to embrace the entire world."[15] Since everything is part of the ego, "Bodies do not produce sensations, but complexes of elements (complexes of sensations) make up bodies." This view serves greatly to simplify the scientific enterprise:

For us, therefore, the world does not consist of mysterious entities, which by their interaction with another, equally mysterious entity, the ego, produce sensations, which alone are accessible. For us, colors, sounds, spaces, times, etc. are provisionally the ultimate elements, whose given connexion it is our

---

[11] Mill (1881), Bk. III, Ch. xii, § 6.
[12] Ibid., § 2.
[13] See e.g. Mach (1906), pp. xii and 29.
[14] From Mach's *Die Geschichte und die Wurzel des Satzes von der Erhaltung der Arbeit*, Prague, 1872; cited in Blackmore (1972), p. 86. See also e.g. Mach (1906), p. 363.
[15] Ibid., p. 13; next quote, p. 29.

business to investigate. It is precisely in this that the exploration of reality consists.[16]

For Mach, reality does not extend beyond what can be known; all that can be known are sensations and their relations; and that unknowable something which purportedly lies behind sensations is the concern of metaphysics. Thus, like Comte, Mach is led to denounce metaphysics.

I should like the scientists to realize that my view eliminates all metaphysical questions indifferently, whether they be only regarded as insoluble at the present moment, or whether they be regarded as meaningless for all time. I should like then, further, to reflect that everything that we can know about the world is necessarily expressed in the sensations, which can be set free from the individual influence of the observer in a precisely definable manner. ... Everything that we can want to know is given by the solution of a problem in mathematical form, by the ascertainment of the functional dependency of the sensational elements on one another. This knowledge exhausts the knowledge of 'reality.'[17]

One might wonder, however, whether Mach's view itself contains metaphysical elements. Not according to Mach: "[O]ur view has no metaphysical background, but corresponds only to the generalized expression of experiences."

Mach does not assume himself to be the first to advocate such a view, and refers to Berkeley and Hume in this context. It also seems that he believes a good deal of science actually to have been conducted in a way which is in keeping with his conception: "Science has always required self-evident propositions as a safe foundation upon which to build;"[18] and he sees his foremost opponents as being those who assume the *existence* of a non-perceptible molecular or atomic realm.

All the same, Mach is apparently not against theorising about such a realm,[19] in that it may lead to the awareness of important relations

---

[16] Ibid., pp. 29–31.

[17] Ibid., p. 46; next quote, p. 61.

[18] Ibid., p. 56.

[19] Though he does say in his (1883), p. 337, that "A thing that is beyond the ken of knowledge, a thing that cannot be exhibited to the senses, has no meaning in natural science."

among phenomena. But once these relations have been determined, the theory and its realm of imperceptibles is to be discarded.[20]

Like Mill, Mach identifies causation with succession; and like both Comte and Mill he conceives of explanation as at most consisting in indicating particular formal relations among phenomena: "a so-called 'causal' explanation, also is nothing more than the statement of an actual fact [sensation] or of a connection between facts."[21] Thus for Mach no non-causal constant relations are possible; and in explanation no reference need be made to unseen agents or forces.

A number of questions arise concerning this view, such as how it is to avoid solipsism, and the extent to which scientific practice actually has been in keeping with it. Concerning this latter point we might consider as an example an aspect of science apparently accepted as legitimate by all empiricists, namely Kepler's three laws of motion – or more particularly the first law – which states that the orbit of each planet is an ellipse, with the sun at one focus.

According to Mach, the value in Kepler's first law lies in its ability to economise our thinking about the motions of the planets. But is this economising of thought consistent with the view that it is fundamentally the phenomena that are 'real'? From a broadly phenomenal point of view the planets do not describe ellipses, but involve retrogradation and so forth. In order to conceive of them as describing ellipses – in order to obtain 'economy of thought' – one must *abstract* from the phenomenal realm.[22] And this abstraction implies a form of conceptualisation in which the phenomenally manifest motions are illusory.

As regards solipsism, one might ask how on Mach's view there can exist anything more than one thinking subject and the sensations experienced by that subject, for the bodies of other people are only

---

[20] In this regard, see Laudan (1981), pp. 211&n–213. A fundamental point with respect to the empiricism/realism issue however is whether such a stance is actually consistent with an otherwise empiricist approach. This point will be returned to below.

[21] Mach (1906), p. 335.

[22] Cf. Whewell (1847), Part 1, p. 25: "When we see a body move, we see it move in a path or *orbit*, but this orbit is not itself seen; it is constructed by the mind."

sensations experienced by the ego, and the sensations experienced by those bodies in turn are not perceived by the ego at all. Mach's reply to the latter point is that each of us is justified in positing the existence of the sensations of others on the basis of analogy with our own case.[23]

Ludwig Boltzmann (1844–1906) uses this line of reasoning against Mach, saying that we can then just as well use analogy to argue for the existence of a non-perceptible realm of atoms.[24] To this Mach has responded by saying: "I never had it in mind to replace the language of the vulgar or even the everyday speech of scientists. Also, I hope I will be credited with having long been familiar with the simple considerations which Boltzmann has raised."[25] Whether Boltzmann's considerations were simple or not, history has refuted Mach on the question of the existence of atoms, in the sense then at issue.

But the basic difference between Boltzmann and Mach lies deeper than this. We can begin to obtain an idea of this difference by comparing Mach's saying that what we represent to ourselves behind the appearances exists only in our understanding, with Boltzmann's disavowal of solipsism:

Solipsism is the view that the world is not real, but a mere product of our [sic] fantasy, like a dream object. I too once hankered after this whim, which led to my failing to take the right practical action and caused me great damage, to my immense delight, for this provided me with the desired proof that the external world exists, a proof that can consist only in showing that if we doubt this existence we are less able to act appropriately.[26]

Here we see that Boltzmann differs from Mach in his belief in the *existence* of a non-phenomenal or real world – a point central to

---

[23] Mach (1906), pp. 24 and 27.

[24] Boltzmann, 'On the Question of the Objective Existence of Happenings in Inanimate Nature,' 1897, in his *Populäre Schriften*, Leipzig, 1925, p. 175; cited in Blackmore (1972), p. 206.

[25] From *Die Principien der Wärmelehre*, Leipzig, 1896; cited in Blackmore (1972), p. 206.

[26] Boltzmann (1900, 1902), pp. 150–151.

understanding the debate between them.[27] In this light we can under-stand why, for example, Boltzmann makes no claim to the effect that the atomic theory must be correct, i.e. that atoms in particular must exist or be real. His position is rather that *something* must be real, and that scientific advance on the phenomenal level requires theorising about the nature of that something. The line he takes is thus that:

All these [previously mentioned] achievements and many earlier attainments of atomic theory are absolutely unattainable by phenomenology or energetics; and I assert that a theory that achieves original insights unobtainable by other means, that is moreover supported by many facts of physics, chemistry and crystallography, such a theory should not be opposed but cultivated further. As regards ideas about the nature of molecules it will however be necessary to leave the widest possible room for manoeuvre.[28]

And we can also understand Mach's position, that atoms *cannot* be realities behind the phenomena, since for him *nothing* can:

[I]t would not become physical science to see in its self-created, changeable, economical tools, molecules and atoms, realities behind phenomena' .... The atom must remain a tool for representing phenomena, like the functions of mathematics. Gradually, however, as the intellect, by contact with its subject matter, grows in discipline, physical science will give up its mosaic play with stones and seek out the boundaries and forms of the bed in which the living stream of phenomena flows.[29]

Where realists claim that there is a reality lying beyond or behind the realm of phenomena, a reality which is at least partly responsi-ble for the nature of that phenomenal realm, strict phenomenalists say that it is quite pointless to postulate even the existence of such a reality, since nothing could be known about it in any case. It would

---

[27] Thus issue is here taken with Larry Laudan, who believes the fundamental difference between Mach and Boltzmann to lie in their divergent *methodological* views with respect to the aim and structure of theory (1981, p. 218). As will be shown here, their differences with regard to theory can more profitably be seen as a manifestation of their more fundamental *ontological* differences – differences which Laudan neglects and which are difficult to imagine being explained on his view.

[28] Boltzmann (1899), p. 99; see also e.g. Philipp Frank's recollection of a conversation with Boltzmann, quoted in Brush (1968), p. 207.

[29] Mach, *Populär-wissenschaftliche Vorlesungen*, Leipzig, 1896; pp. 206–207 in the translation: *Popular Scientific Lectures*, La Salle, Ill., 1943; quoted in Laudan (1981), p. 224.

at best be a noumenal world of *Dinge an sich*, the subject of unend-ing metaphysical speculation. Realists, on the other hand, feel that the only way one can make sense of the world of phenomena is by assuming the existence of just such a real world. They are thus forced to go beyond experience in their considerations, and to treat of what is not certain. In science this means *theorising* about reality, while recognising the tentative nature of one's efforts. In this regard Boltz-mann is perhaps one of the first explicitly to acknowledge the value of the transitory in science, and implicitly to suggest that the attainment of certainty or truth need not be science's sole aim:

[H]ypotheses that leave some play to fantasy and go more boldly beyond what is given will give constant inspiration for novel experiments and thus become pathfinders to totally unsuspected discoveries. Such a theory will of course be subject to change and it may happen that a complicated theoretical structure will collapse and be replaced by a new and more effective one, in which however the old theory as a picture of a restricted field of phenomena usually continues to find a place within the framework of the new one; as for instance emission theory for describing catoptric and dioptric phenom-ena, the hypothesis of an elastic luminous aether for representing interfer-ence and diffraction, the theory of electric fluids for describing electrostatic phenomena.[30]

That the attainment of truth be the primary, if not the only aim of science is a presupposition of virtually all non-realist contributors to the debate. In this respect, **Henri Poincaré** (1854–1912) is no exception: "The search for truth should be the goal of our activities; it is the sole end worthy of them."[31] Poincaré's ontological position is that of a strict positivist or phenomenalist:

[A] reality completely independent of the mind which conceives it, sees or feels it, is an impossibility. A world as exterior as that, even if it existed, would for us be forever inaccessible. But what we call objective reality is, in the last analysis, what is common to many thinking beings, and could be common to us all; this common part, we shall see, can only be the harmony expressed by mathematical laws. It is this harmony then which is the sole objective reality, the only truth we can obtain.

---

[30] Boltzmann (1904), p. 161.
[31] Poincaré (1914), p. 11; next quote, p. 14.

In considering the nature of science in more detail, Poincaré makes a distinction similar to Whewell's distinction between metaphysical maxims, theories of causes, and laws of phenomena. Poincaré's distinction is between three categories of hypothesis: the first category contains general principles, the second what he calls 'indifferent hypotheses,' and the third 'real generalisations,' which are the mathematical laws referred to in the quote above. Unlike Whewell's metaphysical maxims however, Poincaré's fundamental principles are not to be metaphysical but are to be arrived at by the experimental method.[32]

Hypotheses of the second category, like Boltzmann's 'theories,' constitute attempts to conceptualise the reality Poincaré considers inaccessible. With regard to them he says: "There is a second category of hypotheses which I shall qualify as indifferent. In most questions the analyst assumes, at the beginning of his calculations, either that matter is continuous, or the reverse, that it is formed of atoms. In either case, his results would have been the same. On the atomic supposition he has a little more difficulty in obtaining them – that is all."[33]

One might wonder whether Poincaré is correct in his assumption that all rival hypotheses (theories) which might fall into his second category necessarily are 'indifferent,' or, in modern terminology, necessarily are 'underdetermined by the data.' Almost anticipating Poincaré's very words, Whewell says:

If any one holds the adoption of one or other of these theories [of emission and undulation] to be indifferent, let him express the *laws of phenomena* of diffraction in terms of the theory of emission. If any one rejects the doctrine of undulation, let him point out some other way of connecting double refraction with polarization. And surely no man of science will contend that the beautiful branch of science which refers to that connexion is not a portion of our positive knowledge.[34]

---

[32] See Poincaré (1902), pp. 105, 110, 129, 138, 166. Poincaré's general principles are more similar to Whewell's empirical specifications of metaphysical maxims and to what in the present study are termed *refined* principles, while Whewell's metaphysical maxims are more similar to what we term *fundamental* principles.

[33] Ibid., p. 152.

[34] Whewell (1860), p. 231.

Like Mach (and Comte in places), Poincaré admits that theories or 'indifferent hypotheses' may have a place in science, in that they may give satisfaction to the mind;[35] but also in the spirit of Mach he warns against their being taken too seriously: "These indifferent hypotheses are never dangerous provided their characters are not misunderstood. They may be useful, either as artifices for calculation, or to assist our understanding by concrete images, to fix the ideas, as we say. They need not therefore be rejected." At another point he adds: "These hypotheses play but a secondary rôle. They may be sacrificed, and the sole reason this is not generally done is that it would involve a certain loss of lucidity in the explanation [*exposition*]."[36] Elsewhere, though, he considers them to be "useless and unverifiable."

We find an expression of Poincaré's *instrumentalism* in the context of a discussion of Maxwell's electrodynamic theory, where he says: "Two contradictory theories, provided they are kept from overlapping, and that we do not look to find in them the explanation of things, may, in fact, be very useful instruments of research."[37]

In his philosophy of science Poincaré not only believes that we cannot have knowledge of real objects,[38] but considers questions regarding their nature to be meaningless. "To those who feel that we are going too far in our limitations of the domain accessible to the scientist, I reply: These questions which we forbid you to investigate, and which you so regret, are not only insoluble, they are illusory and devoid of meaning." On the other hand, the true relations *between* real objects *can* be known.[39] According to Poincaré these relations take the form of mathematical (empirical) laws, and not scientific theories; and for him it is to the attainment of the knowledge of such laws that science ought ultimately be devoted. "The day will perhaps come when physicists will no longer concern themselves with questions which are inaccessible to positive methods,

---

[35] Poincaré (1902), p. 164; quote following, p. 153.
[36] Ibid., p. 212; next quote, p. 156.
[37] Ibid., p. 216; for the Perspectivist conception of theory conflict and the realist-instrumentalist controversy, see Dilworth (2007), pp. 105–107.
[38] Poincaré (1902), p. 161; quote following, p. 163.
[39] Ibid., p. 161; quote following, p. 223.

and will leave them to the metaphysicians. That day has not yet come; man does not so easily resign himself to remaining forever ignorant of the causes [*le fond*] of things."

Pierre Duhem (1861–1916) distinguishes between two kinds of theory: what may be called *explanatory* theories, which are similar to Whewell's theories of causes and Poincaré's indifferent hypotheses, and what might be termed *representational* theories, which are systematisations of experimental laws and are similar to Poincaré's hypotheses containing general principles.[40] Where Comte, Mill and Mach would call systematisation explanation, Duhem keeps the two cases separate. According to Duhem, "While we regard a physical theory as a hypothetical explanation of material reality, we make it dependent on metaphysics. In that way, far from giving it a form to which the greatest number of minds can give their assent, we limit its acceptance to those who acknowledge the philosophy it insists on."[41]

Where Poincaré sees the search for truth as the sole aim worthy of science, Duhem's view is similar, though he expresses this aim as a striving to obtain universal assent. He therefore sees his own task as being to provide a conception of theories which allows of their being impartially judged, i.e. judged independently of metaphysical bias. On Duhem's conception, "A physical theory is not an explanation. It is a system of mathematical propositions, deduced from a small number of principles, which aim to represent as simply, as completely, and as exactly as possible a set of experimental laws."[42]

Thus for Duhem,

a true theory is not a theory which gives an explanation of physical appearances in conformity with reality; it is a theory which represents in a satisfac-

---

[40] In this regard see Giedymin (1982), p. 79: "The distinction between these two types of theory (and theorists) had in fact been made by Poincaré ... in his 1888–9 lectures on electricity and optics" – cf. the Introduction to Poincaré's *Electricité et optique: les theories de Maxwell et la théorie électromagnetique de la lumière*, J. Blondin (ed.), Paris: G. Carré, 1890.

[41] Duhem (1906), p. 19. Often in this work Duhem speaks as though he were advocating a particular way of viewing *any* physical theory whether it be explanatory or representational, when in fact he is advocating a way of theorising which excludes the former.

[42] Ibid., p. 19; next quote, pp. 20–21, 26–27.

tory manner a group of experimental laws. A false theory is not an attempt at an explanation based on assumptions contrary to reality; it is a group of propositions which do not agree with the experimental laws. *Agreement with experiment is the sole criterion of truth for a physical theory.* ...

Thus, physical theory never gives us the explanation of empirical laws; it never reveals realities hiding under the sensible appearances; but the more complex it becomes, the more we apprehend that the logical order in which theory orders experimental laws is the reflection of an ontological order, the more we suspect that the relations it establishes among the data of observation correspond to real relations among things, and the more we feel that theory tends to be a natural classification.[43]

Duhem thus takes issue both with the use of picturable mechanical models, and with atomistic theories. The former satisfy only those with a "need to imagine concrete, material, visible, and tangible things." In this context Duhem also shows an aversion to a particular form of instrumentalism. For the tangibly inclined, theory is "neither an explanation nor a rational classification of physical laws, but a model of these laws, a model not built for the satisfying of reason but for the pleasure of the imagination. Hence, it escapes the domination of logic. ... To a physicist of the school of Thomson or Maxwell, there is no contradiction in the fact that the same law can be represented by two different models."[44]

Atomistic theories, for their part, presuppose the metaphysics of atomism:

[E]ach time the fortunate daring of an experimenter discovers a new set of experimental laws, he will see the atomists, with feverish haste, take possession of this scarcely explored domain and construct a mechanism approximately representing these new findings. Then, as the experimenter's discoveries become more numerous and detailed, he will see the atomist's combinations get complicated, disturbed, overburdened with arbitrary complications without succeeding, however, in rendering a precise account of the new laws or in connecting them solidly to the old laws; and during this period he will see abstract theory, matured through patient labor, take possession of the new lands the experimenters have explored, organize these con-

---

[43] Ibid., p. 70.

[44] Ibid., p. 81; with reference to Poincaré, cf. also Duhem (1905), p. 294. Duhem's conception of models here corresponds to what will be treated as *analogues* beginning in Chapter 5 of the present work.

quests, annex them to its old domains, and make a perfectly coordinated empire of their union.[45]

Duhem's position leads him to criticise Galileo, whom he terms a realist, for suggesting that the earth *really* revolves about the sun, a view which transgresses Duhem's injunction not to take theories as telling us anything about reality. Duhem sides rather with Cardinal Bellarmine, whom he quotes: "It is one thing to prove that by assuming the sun at the center of the world and the earth in the heavens one saves all the appearances, and quite another thing to demonstrate that the sun really is at the center, the earth really in the heavens."[46]

Galileo's reply to this line of reasoning is:

Granted, it is not the same thing to show that on the assumption of the sun's fixity and the earth's mobility [all?] the appearances are saved and to demonstrate that such hypotheses are really true in nature. But it should also be granted, and is much more true, that on the commonly accepted system there is no accounting for these appearances, whence this system is indubitably false; just so should it be granted that a system that agrees very closely with appearances may be true; and one neither can nor should look for other or greater truth in a theory than this, that it answers to all [*sic*] the particular appearances.[47]

Duhem believes Galileo's reasoning here actually to lead to the view Duhem himself endorses. It seems however that Galileo, unlike Duhem, does not intend that our knowledge cannot extend beyond the phenomenal realm, but that it is in precisely the manner here

---

[45] Ibid., p. 304.

[46] Excerpted from letter of Bellarmine to Foscarini, dated April 12, 1615; pp. 121–125 in *Copernico e le vicende del sistema copernicano in Italia nella seconda metà del secolo XVI e nella prima del secolo XVII*, Rome, 1876; p. 107 in Duhem (1908a). Note that the issue here is not over the existence of 'theoretical' entities, but over two conceptions which are being considered only with regard to their kinematic differences.

[47] *Copernico* ... (op. cit.), pp. 129–130; pp. 108–109 in Duhem (1908a). The notion of saving the phenomena originates with Plato and is almost always used in astronomical contexts. In Plutarch, for example, Cleanthes is depicted as thinking "that the Greeks ought to lay an action for impiety against Aristarchus the Samian on the ground that he was disturbing the hearth of the universe because he sought to save [the] phenomena by assuming that the heaven is at rest while the earth is revolving along the ecliptic and at the same time is rotating about its own axis." (Plutarch, p. 55).

favoured by Duhem that it does just that. Both Galileo and Duhem admit the *existence* of a non-phenomenal reality, and Duhem in the present context allows that we can *assume* it to be of a particular nature if by so doing the phenomena can be saved. Galileo says that our assumption may be true, which Duhem cannot deny, and he may be suggesting that the sole criterion for its being so is whether it saves *all* the phenomena.

These two views are very similar, and one would expect that the methodological consequences of adopting either would differ little from those of adopting the other. The question remains, however, whether the method of theorising advocated by Duhem would naturally lead to the making of assumptions of the sort at issue here, and whether the making of such assumptions, if only hypothetically, is not itself a form of realism.[48]

But then Duhem has himself expressed views which seem more sympathetic to realism than to positivism:

If, on the other hand, [the physicist] yields to the nature of the human mind, which is repugnant to the extreme demands of positivism, he will want to know the reason for, or explanation of, what carries him along; he will break through the wall at which the procedures of physics stop, helpless, and he will make an affirmation which these procedures do not justify; he will be metaphysical.

What is this metaphysical affirmation that the physicist will make, despite the nearly forced restraint imposed on the method he customarily uses? He will affirm that underneath the observable data, the only data accessible to his methods of study, are hidden realities whose essence cannot be grasped by these same methods, and that these realities are arranged in a certain order which physical science cannot directly contemplate.[49]

And though Duhem's original aim was to provide a conception of theory free from metaphysics, at the end of the appendix to his main work in the philosophy of science he gives expression to a conception reminiscent of Whewell: "[T]he physicist is compelled to recognize that it would be unreasonable to work for the progress of physical theory if this theory were not the increasingly better defined and more

---

[48] We shall return to this issue later in the present chapter when we take up the views of Bas van Fraassen.

[49] Duhem (1905), pp. 296–297.

precise reflection of a metaphysics; the belief in an order transcending physics is the sole justification of physical theory."[50]

N. R. Campbell (1880–1949), like Duhem, considers the attaining of truth to be an important aspect of science, and the universal provision of assent to constitute the criterion of whether a judgement is part of the subject-matter of science.[51] But he attributes a different significance to this assent than does Duhem, taking it not to concern the realm of phenomena, but an independent reality. He says: "of course scientific reality must be independent of perception in a certain degree. It must be independent of the perceiver. This is what we mean when we say that there must be universal agreement concerning the laws which define reality." And with respect to Berkeley, considered e.g. by Mach to be a "far more logically consistent thinker than Kant,"[52] Campbell says: "Now his view of reality seems exactly contrary to that we have just considered; reality, so far from being independent of perception, consists of nothing else than its being perceived. To me, and I believe to most men of science, the idea is unintelligible."[53]

Campbell also largely removes himself from the view of Mill, attacking for example his (Humean) conception of causality as succession, pointing out that common sense shows it to be too wide:

Thus, there is the hackneyed instance of the relation of day and night; each day invariably precedes the following night, and yet we do not regard that day as the cause of that night. Or, to take an even simpler example, if we allow a body to fall freely, its fall through the first foot precedes invariably its fall through the second; and yet the fall through the first foot is not usually regarded as the cause of the fall through the second. Every fact which precedes invariably another fact is not its cause.

Further, unlike Duhem, Mach, Poincaré and others of a phenomenalist persuasion, Campbell does not consider truth to be science's sole or ultimate aim. As he sees it, "The search for truth alone never has and never will lead to any science of value. The spirit which must be so carefully curbed in the search for truth must be given free

---

[50] Duhem (1908b), p. 335.
[51] Campbell (1920), p. 21; quote following, p. 254.
[52] Mach (1906), p. 368.
[53] Campbell (1920), p. 255; next quote, p. 58.

rein when truth is attained."[54] What is required apart from truth is
what Campbell terms *meaning*. He says: "a proposition is true in so
far as it states something for which the universal assent of all mankind
can be obtained; it has meaning in so far as it gives rise to ideas which
cause intellectual satisfaction."

In this context Campbell distinguishes between laws and theories,
saying that "The value of a law lies chiefly in its truth, that of a theory chiefly in its meaning."[55] Laws, he believes, are almost always
suggested by theories,[56] which are of two kinds – 'mechanical' and
'mathematical' – corresponding to Duhem's explanatory and representational theories.[57] Campbell however considers both kinds to be
explanatory – to provide intellectual satisfaction – not of individual phenomena, but of laws, it being "the business of science not
only to discover laws but also to explain them." Mechanical theories accomplish this by indicating that the laws to be explained are
analogous to other laws with which we are more *familiar*; and mathematical theories by unifying them under a more general proposition,
thereby *simplifying* them. Campbell suggests however that this latter form of explanation might actually be better termed generalisation,[58] and that "For those to whom theories of the first type are of
supreme importance, the importance of those of the second type consists chiefly in the prospect of their ultimate conversion to those of the
first."

Where Duhem's distinction between explanatory and representational theories is clearly one between realist and phenomenalist
approaches to science, Campbell's distinction is less evidently so. That
he has something like this in mind, however, shows itself in his consideration of the relative values of 'mechanical' and 'mathematical'
theories. Here he mentions such points as that mechanical theories,
even if false, have nevertheless stimulated research, and that though

---

[54] Ibid., p. 224; next quote, p. 218.
[55] Ibid., p. 219. We recall here Poincaré's allowing that 'indifferent hypotheses'
(theories) might give satisfaction to the mind, while considering this to be of little
consequence to science.
[56] Ibid., pp. 88, 104–105.
[57] Cf. ibid., pp. 150n., 151n.; quote following, p. 113.
[58] Ibid., p. 146; quote following, p. 148.

they may be more liable to error than mathematical theories, this can be excused by the fact that they state more. He also implies that a generalisation based solely on the phenomena to be explained does not provide those phenomena with the certainty they would receive were they explained by a conceptually independent theory. Thus:

The test of 'purely phenomenal' gives exactly the opposite result to that imagined by its proposers; the more purely phenomenal a proposition is and the less the element of theory associated with it, the less is its certainty. From phenomena and phenomena alone we can deduce only negative conclusions, never a positive; it is theory which gives us positive certainty.[59]

While admitting that "A taste for mechanical theories can no more be forced by argument on one who does not possess it than a taste for oysters," Campbell considers what might lie behind the intellectual satisfaction so many scientists find in mechanical theories. Taking what he considers to be a paradigmatic case, in which the laws to which analogy is made in a mechanical theory are Newton's laws of motion, he suggests that the satisfaction provided by the theory stems mainly from its involving the notion of bodies moving under the influence of forces, which we can each of us relate to our own voluntary actions of causing physical objects to move.[60] Unlike most authors, however, Campbell does not present the issue as though the decision lies solely with us, but grants a role also to nature, recognis-

---

[59] Ibid., p. 153; next quote, p. 151.

[60] Ibid., p. 155; see also pp. 63–64. This view is also expressed by Whewell, who examines the etymology of the term "force": "The original meaning of the Greek word [for *Force*] was a *muscle* or *tendon*. Its first application as an abstract term is accordingly to muscular force:

Then Ajax a far heavier stone upheaved, He whirled it, and impressing Force intense Upon the mass, dismist it.

"The property by which bodies affect each other's motions, was naturally likened to that energy which we exert upon them with similar effect: and thus the labouring horse, the rushing torrent, the descending weight, the elastic bow, were said to exert force. ...

"Thus man's general notion of force was probably first suggested by his muscular exertions. ..." (1847), Part 1, pp. 185–186.

ing that "It is possible that in the future we shall have to choose between the advantages of simplicity and those of familiarity."[61]

In keeping with the view of Whewell,[62] and in direct opposition to Mill, Campbell sees formal logic as being of no relevance to science.

Of course the province and power of logic have been very greatly extended in recent years, but some of its essential features... have remained unchanged; and any process of thought which does not show those features is still illogical. But illogical is not synonymous with erroneous. I believe that all important scientific thought is illogical, and that we shall be led into nothing but error if we try to force scientific reasoning into the forms prescribed by logical canons. [S]cientific thought is fundamentally different from logical thought.[63]

Campbell supports his conviction by suggesting that where for logicians *words* or equivalent symbols are the instruments by means of which the process of thought is conducted, in science words have nothing to do with the operation whereby one passes from one set of thoughts to another. For the scientist, thoughts are primary, and words are simply a convenient means of calling thoughts to mind. As regards e.g. the idea of providing formal definitions, for example one of *silver*:

No student of science has ever felt the smallest need for a formal definition of silver; our words are perfectly effective in calling up the thoughts we desire without one, and in admitting the right of anyone to ask for one we are encouraging a very dangerous delusion. ... When we are merely trying to call up ideas, a jumble of words quite outside all the rules of grammar may often be more effective than the most accurately turned sentence.[64]

And, for example, whether two statements call up essentially similar ideas cannot be ascertained by studying their grammatical form. In summary, Campbell concludes that "the reasoning processes of classical logic have no application to science."

---

[61] Campbell (1920), p. 157.
[62] See e.g. p. 67, n. 31, of the present work.
[63] Campbell (1920), p. 52. This, together with the rest of Campbell's text cited here, should dispel the myth that Campbell is one of the fathers of logical empiricism.
[64] Ibid., p. 53; next quote, p. 38.

Quite the contrary of this view was adopted by the more influential members of the **Vienna Circle** and their successors, who wrote as though they believed science to be essentially a linguistic phenomenon, and all epistemological relations to be between sentences in 'the language of science.' In conjunction with this, according to their programme of logical *positivism* – later to evolve into logical *empiricism*,[65] the subject-matter of science was to consist of what can be directly observed. Thus, theoretically, the 'language of science' was to consist of logical operators and terms referring to what is directly observable, while practically, it was taken to be that part of everyday language capable of being represented in terms of first-and second-order predicate logic.

This basic view was so influential that even those who considered themselves the severest critics of logical empiricism, such as Karl Popper and Imre Lakatos, did not challenge the programme as regards these fundamentals.[66] In fact, the view has been so widely held during the twentieth century as practically to constitute the discipline of the philosophy of science itself; and its influence on those who believe themselves to have moved beyond it is still strong today, as will be seen below.

What is remarkable from the point of view of the survey of the present chapter is the dogmatic acceptance of this approach in the early work of such of its proponents as **Rudolf Carnap**, **Ernest Nagel** and **Carl Hempel**, who not only provide no argument for it, but tacitly assume even theoretical science, as actually practised, to be in keeping with it. We shall here look at the work of **Hempel**, in which themes of relevance to the empiricism/realism issue are most clearly developed.

Hempel is perhaps best known in connection with the deductive-nomological model of scientific explanation, his general account of which he admits, citing Mill and others, to be 'by no means novel.' Though he remarks in his text that there exists a method of explaining phenomena using e.g. micro-theories, thereby obtaining insight into certain inner mechanisms, he believes this method to be

---

[65] On the conceptual development of logical empiricism, see Dilworth (2007).

[66] In this regard, see ibid.

but an instance covered by his own conception.[67] Apparently unaware of the existence of an issue in this regard, he assumes the
extreme empiricist stance that causality is captured by statements
of empirical regularity, that (causal) explanation in actual science
consists in nothing other than systematisation under such statements, and that explanation and prediction have the same (syllogistic)
form.[68]

Thus on Hempel's way of thinking, "Scientific laws and theories
have the function of establishing systematic connections among the
data of our experience, so as to make possible the derivation of some
of those data from others." Some ten years after making this claim,
however, he wonders: "Why should science resort to the assumption
of hypothetical entities when it is interested in establishing predictive and explanatory connections among observables?"[69] And after
lengthy considerations in terms of elementary and second-order logic,
he concludes that it is not in fact the sole purpose of scientific theory to establish deductive connections among observation sentences
– though he does not tell us what other purpose he thinks it might
have.

Later in his career, when he recognises an alternative to his original view, Hempel argues against what he believes this alternative
to be. While admitting that actual explanations of working scientists
are not in keeping with his covering-law model, he claims this to
be due to their being formulated with a particular kind of audience
in mind. Thus, according to Hempel, such explanations are not objective, and consequently their existence does not suffice to show
his model of (audience-independent) explanation to be "hopelessly
inadequate."[70]

In a similar vein he criticises the idea that explanation involves
what he terms a 'reduction' to the familiar, since: "what is familiar
to one person may not be so to another." Furthermore, "instead
of reducing the unfamiliar to the familiar, a scientific explanation
will often do the opposite: it will explain familiar phenomena with the

---

[67] Hempel & Oppenheim (1948), p. 259.
[68] Ibid., pp. 246–251; quote following, p. 278.
[69] Hempel (1958), p. 179.
[70] Hempel (1965), pp. 427–428; next quote, p. 430.

help of theoretical conceptions which may seem [*sic*] unfamiliar and even counter-intuitive, but which account for a wide variety of facts and are well supported by the results of scientific tests."[71] However, as regards what should constitute an adequate scientific explanation, this just begs the question. So too does Hempel's stating, without drawing an essential distinction between models and analogies, that "For the systematic purposes of scientific explanation, reliance on analogies is . . . inessential and can always be dispensed with."

Though in still later work Hempel does come to admit that models as distinct from analogies are of more than heuristic value,[72] and that his conception of objective cognitive significance is much too restrictive,[73] he never makes the step to realising that his problems of scientific explanation and of theoretical entities are but manifestations of a deeper philosophical issue concerning empiricism and realism. This deeper issue is of central concern, however, to **Rom Harré**, who reacts strongly to the underlying presuppositions of logical empiricism, and develops a comprehensive realist conception of science in its stead.

The empiricists and Harré both believe that there is nothing fundamentally wrong with the way science is actually practised. But for Harré actual scientific practice is more important than it is for the empiricists, and he stresses that "the philosophy of science must be related to what scientists actually do, and how they actually think."[74]

In this light Harré, like Whewell and Campbell, takes the empiricists to task for their unquestioned assumption that formal logic provides the appropriate tools for the analysis of science: "Traditional logic recognises only truth and falsity, and these are very peripheral concepts in science. We must not fall into thinking that an intellectual process, like the process by which theories are formulated, is not capable of an analysis to reveal its rationale, just because the simple principles of truth and falsity cannot capture it."[75]

---

[71] Ibid., p. 431; next quote, p. 439.
[72] Hempel (1970), pp. 157–158.
[73] Hempel (1973), p. 377.
[74] Harré (1972), p. 29.
[75] Harré (1970a), p. 281. Wesley Salmon advances a similar view: "attempting to give a logical characterization of scientific explanation is a futile venture, and . . .

Harré questions not only the notions of truth and falsity as engendered by logic, but also the formalistic idea that science aims to establish deductive relations among statements.

If it is maintained that criticisms of deductivism which hinge on the divergence between that theory and practice are beside the point since the theory depicts an ideal form of knowledge, then deductivism as an ideal should be free from 'problems.' But it is just the very theory which runs fastest into difficulties. To hold that [the idea that] scientific knowledge should develop towards a deductively organized system of conditional statements, describing regularities of succession among types of events, is the *logical* ideal, leads us straight into a situation in which we have to say that in their ideal form scientific theories are not confirmable (Problem of Induction), are about regularities of sequences of types of events indistinguishable from accidental sequences (Problem of Causality), and change their plausibility relative to evidence under purely logical transformation (Problem of Instance Confirmation), and that scientific laws are indistinguishable logically from accidental generalizations (Problem of Natural Necessity), and are such that their component predicates are logically independent (Problem of Subjunctive Conditionals).[76]

Rather than aiming to systematise descriptions of phenomena, according to Harré, "science *above all* seeks to answer the question *why?* ... The main problem of science is to discover by what means the non-random patterns in nature are produced."[77] This involves more than the determining of empirical correlations.

It is hardly a scientific explanation of phenomena merely to describe some other phenomena with which they are associated, unless one has some conception of how this association comes about. Then that conception is really what is doing the explaining and is the heart of the theory. For instance, it is not a scientific explanation of the Aurora to instance the increased activity of sun-spots which regularly antedate the appearance of the glow in the sky. A scientific explanation will tell you why and how the sun-spots are associated with the Aurora, and this involves discussions of the nature of sun-spots and of the paths of electrons which leave the sun. These discussions are relevant only because we have some idea about the nature of the Aurora,

---

little of significance can be said about scientific explanation in purely syntactical or semantical terms." (1984, p. 240).

[76] Harré (1970b), pp. 28–29.

[77] Harré (1970a), p. 279.

and know a good deal about the discharge of electricity in tenuous gases. In short, to explain the Aurora we describe the mechanism which produces the phenomenon, and so come to see *why* sun-spots are associated with the Aurora.[78]

At another point Harré clarifies what he means by the term "mechanism":

In ordinary English this word has two distinct meanings. Sometimes it means mechanical contrivance, a device that works with rigid connections, like levers, the intermeshing teeth of gears, axles, and strings. Sometimes it means something much more general, namely any kind of connection through which causes are effective. It is in the latter sense that I mean the word, [and it] is in the latter sense that the word is used in science generally, in such diverse expressions as the mechanism of the distribution of seeds and the mechanism of star formation.

And, for Harré, mechanisms are depicted by *models*:

Scientists, in much of their theoretical activity, are trying to form a picture of the mechanisms of nature which are responsible for the phenomena we observe. The chief means by which this is done is by the making or imagining of models. Since enduring structures are at least as important a feature of nature as the flux of events, there is always the chance that some models can be supposed to be hypothetical mechanisms, and that these hypothetical mechanisms are identical with real natural structures.[79]

Following this line, Harré also suggests prediction and explanation in actual science to be more fundamentally different than is recognised on the logical empiricist view. Sometimes prediction is possible where explanation is not, and vice versa:

Consider the course of a disease. Long before any explanation of what happens is available the empirical knowledge of experienced doctors may enable them to foretell the course of the disease with great accuracy from the symptoms. We would hardly call the description of the symptoms the explanation of the later stages of the disease. Nor indeed would we call the predictions made from nautical almanacs the explanation of the risings, settings, and conjunctions of the heavenly bodies, Characteristically, giving an explanation involves describing the mechanism, usually the causal mecha-

---

[78] Harré (1972), p. 24; next quote, p. 118.
[79] Harré (1970b), pp. 34–35.

nism, responsible for a series of happenings, and this may not be enough to predict just what will happen. We know the causal mechanism of evolutionary change pretty well, but until we actually observe what happens we are unable to predict the appearance of new forms of plants and animals, because of the presence of the random (unpredictable) element of mutation in the system.[80]

So, for Harré, providing a scientific explanation consists essentially in depicting a relevant causal mechanism by means of a model. But the empiricist can respond to this approach by saying that the analysis it provides has not gone any deeper than that in terms of constant conjunction, this notion simply reappearing in the conception of the functioning of the causal mechanism.

Note that such a criticism does not question that scientific investigation is actually conducted in the way depicted by Harré, and if we grant that his characterisation is correct, then the criticism should be directed at scientists themselves. But since Harré accepts his conception of science as also being that of the best way to go about investigating the world, he might be expected to defend it on this point.

Harré does provide a reply to the empiricist here, namely that the acceptability of the depiction of a causal mechanism is not dependent on its suggesting a constant conjunction, but in there being "a pair of components, in [the] mechanism, which are related by the basic interactions of the science we are using. Any scientific enquiry with which the reader is familiar can provide instances of this procedure, e.g. physics supplies the mechanisms for chemical phenomena, chemistry for many physiological phenomena, etc., etc."[81]

Harré's realist conception has received relatively little attention in mainstream philosophy of science, in spite of the fact that the empiricism/realism issue has once more become a topic of popular concern. **Bas van Fraassen**, for example, an influential contributor to the current debate, takes no notice of Harré's views in his main work devoted to the issue. In it van Fraassen argues for what he terms 'constructive empiricism,' according to which:

---

[80] Harré (1972), p. 56.
[81] Harré (1970b), pp. 109–110.

Science aims to give us theories which are empirically adequate; and acceptance of a theory involves as belief only that it is empirically adequate ... with the preliminary explication that a theory is empirically adequate exactly if what it says about the observable things and events in this world, is true – exactly if it 'saves the phenomena.'[82]

The implication here is that what the theory 'says' about *un*observable things and events need not be true, and so a distinction must be made between what is observable and what is not. In this regard van Fraassen claims that though "'observable' is a *vague predicate*,"[83] the distinction can nevertheless be drawn. "A look through a telescope at the moons of Jupiter seems to me a clear case of observation, since astronauts will no doubt be able to see them as well from close up. But the purported observation of micro-particles in a cloud chamber seems to me a clearly different case."

He contrasts the above position with what he calls scientific realism, namely the view that "Science aims to give us, in its theories, a literally true story of what the world is like; and acceptance of a scientific theory involves the belief that it is true."[84]

Each of the views depicted by van Fraassen has two parts, one concerning the aim of science and the other concerning people's beliefs. The first is apparently intended as a factual claim about science – "The question is what aim scientific activity has"[85] – while the second seems more prescriptive, and involves the (counter-intuitive) assumption that we can *decide* "how much we shall believe when we accept a scientific theory."

Judging from his terminology in his depiction of constructive empiricism, it would appear that van Fraassen wishes to defend a position similar to Duhem's with regard to the Bellarmine-Galileo dispute. That is, he would admit that it is legitimate to construct theories referring to what transcends experience, but that the correctness of such a theory with respect to empirical matters would not imply its correctness with respect to the realm beyond experience.

---

[82] van Fraassen (1980), p. 12. As in the case of Duhem, it is not altogether clear here how 'saving' the phenomena is to differ from hypothetically explaining them.

[83] Ibid., p. 16; next quote, pp. 16–17.

[84] Ibid., p. 8.

[85] Ibid., p. 18; next quote, ibid.

There is a difference between van Fraassen's and Duhem's views however which is hinted at in the former's suggestion that 'observability' is a vague predicate. Unlike Duhem, Mach, and other phenomenalists, who claim that all that can be known are phenomena (and the relations between them), van Fraassen adopts a form of *naive realism* which admits that we can know of the existence of actual physical objects if they are observable. This means that we must distinguish here between the trans-empirical or real world, distinguished from the phenomenal world as a matter of principle, and van Fraassen's unobservable world, which is neither more nor less real that his observable world, and is distinguished from it only contingently. And we note that this is what leads van Fraassen to suggest that the notion of observability is essentially vague.[86]

But, even setting this realist element in van Fraassen's thinking aside, as intimated earlier with regard to Duhem, the very idea of 'saving the phenomena,' in that it admits the *relevance* of reference to what is beyond experience, is actually more a form of realism than of empiricism. Furthermore, as regards van Fraassen's depiction in particular, what a theory referring to unobservables 'says about observable things and events' is that they take the form they do as a consequence of the nature and behaviour of the unobservables. So his own manner of expression suggests that a theory's being 'empirically adequate' implies its also being 'theoretically adequate,' i.e. that it correctly depict what is unobservable.

Not only is van Fraassen's 'constructive empiricism' actually a form of realism, but his 'scientific realism' is really a form of empiricism. He claims that the realist considers the aim of science to be to provide 'a literally true story of what the world is like,' i.e. to provide a correct description. But as Duhem recognised in distinguishing between explanatory and representational theories, for the realist the aim of science, as far as its theoretical aspect is con-

---

[86] On this point, cf. Ernan McMullin (1984a), p. 20: "Why could there not, in principle, be organisms much smaller than we, able to perceive microentities that for us are theoretical and able also to communicate with us? Is not the notion 'observable in principle' hopelessly vague in the face of this sort of objection? How can it be used to draw a usable distinction between theoretical entities that do have ontological status and those that do not?"

cerned, is primarily to explain and not to describe. It is the empiricist who sees truth as the sole or primary aim of science, and who wants to limit 'theorising' to the establishment of the non-controversial, representational type of theories where he believes such truth can be attained.[87] The realist, on the other hand, in attempting to explain, is forced to speculate. But it is to be admitted that in order for the realist's proposed explanation actually to be an explanation, the trans-empirical realm depicted by his or her theory should *in essence* be as the theory suggests it to be; and to this extent the realist aims to have theories which correctly depict the trans-empirical. But for the realist, unlike the empiricist, 'truth' is not the sole aim of science, and a theory which is not *known* to be correct will be accepted, so long as there is no reason to believe it to be e.g. essentially incorrect, or nonsensical, or incompatible with some other theory the realist accepts.[88]

van Fraassen's characterisations of 'constructive empiricism' and 'scientific realism' thus indicate that he has failed to come to grips with the real issue. Perhaps most notable in this regard is his lack of awareness of the fundamental idea that the primary purpose of constructing theories which depict unobservable entities is not to de-

---

[87] It is of some interest to note here that, as has been pointed out by Brian Ellis (1985, pp. 55, 56) and admitted by van Fraassen (1985, pp. 275, 287), the latter's view originates with studies of representational or systemic theories as versus causal or explanatory theories. On having this be pointed out, however, van Fraassen, rather than consider whether there may be epistemologically important differences between the two kinds of theory, instead takes issue with the (metaphysical) notion of causality. Not seeing it as expressing anything more than asymmetric regularity, he believes it can be eliminated; and in the case of science he asserts that to see causal theories in the same way as representational theories "gives us the best hope of eliminating metaphysics from our interpretation of science" (ibid., p. 288), as though this were a goal shared by parties on both sides of the debate.

[88] On this point cf. McMullin, who supports a realist view: "I do not think that acceptance of a scientific theory involves the belief that it is true. Science aims at fruitful metaphor and at ever more detailed structure. To suppose that a theory is literally true would imply, among other things, that no further anomaly could, in principle, arise from any quarter in regard to it. At best, it is hard to see this as anything more than an idealized 'horizon-claim,' which would be quite misleading if applied to the actual work of the scientist." (1984a, p. 35).

scribe those entities, but to explain observable phenomena.[89] In fact, we might say that it is just this distinction that van Fraassen is himself groping towards in his characterisations of constructive empiricism and scientific realism, the former emphasising the explanatory function of a theory, and the latter the theory's ability to provide a correct description of the world.

A perusal of van Fraassen's work shows that the basic reason for this misunderstanding is that his thought is wholly restricted to the conceptual framework of logical empiricism. This is revealingly expressed, for example, in his liberal and unreflecting use of the notion of 'the language of science,'[90] a notion which arose with logical empiricism, and whose viability is far from evident to anyone critical of its fundamental presuppositions. But more substantially, his depiction of realism primarily in terms of truth is realism as it might be conceived *within* logical empiricism, where the expressions of both laws and theories are assumed to be statements having 'truth-values.' Further, van Fraassen himself conceives of theories in this way throughout, without noting that this conception is highly problematic with respect to the realism/empiricism issue.[91] Nor, like the early Hempel, does he realise that the notions of cause and explanation are similarly problematic; for though he treats of them, he does not touch on the question central to the debate, namely whether the empirical is to be conceived of as being causally based in the trans-

---

[89] This point has been widely missed, even by those advocating various forms of realism. Thus with regard to 'truth,' for example, even if we were to describe realism as involving a search for true theories, the truth of such theories would not lie simply in their providing correct descriptions of the trans-empirical, but in their providing correct descriptions of how the empirical is a manifestation of the trans-empirical.

[90] Cf. e.g.: "Not every philosophical position concerning science which insists on a literal construal of the language of science is a realist position." van Fraassen (1980), p. 11. It may be noted that Whewell also uses this expression, though not in the all-embracing logical empiricist sense, but rather in referring to the technical terms of science.

[91] Cf. again McMullin: "The realist would not use the term 'true' to describe a good theory. He would suppose that the structures of the theory give some insight into the structures of the world. But he could not, in general, say how good the insight is. He has no independent access to the world, as the antirealist constantly reminds him." (1984a, p. 35).

empirical, or whether scientific explanation requires reference to such a realm.

Other contemporary writers, such as **Larry Laudan**, have had similar difficulty in coming to grips with the issue. According to Laudan "the realist maintains that the goal of science is to find ever truer theories about the natural world;"[92] and "Realists in particular argue that scientists should and do seek true theories and, moreover, that they often find them (or at least close approximations to them)." At another place he criticises what he considers to be a form of realism according to which, among other things, "Scientific theories (at least in the 'mature' sciences) are typically approximately true, and more recent theories are closer to the truth than older theories in the same domain."[93]

Here too theories are conceived along logical empiricist lines to be entities which are either true or false, and whose function is to describe rather than explain; and Laudan's reliance on this tradition is also manifest in his frequent use of such turns of phrase as "any language rich enough to contain negation,"[94] and "even highly successful theories may well have central terms that are nonreferring." So, much of what has been said above with regard to van Fraassen can also be said of Laudan, including that his arguments are against realism as it might be seen *within* the context of logical empiricism, and that they in fact support realism.

We see this distortion when we look, for example, at Laudan's discussion regarding theories which have largely had correct empirical consequences, but whose trans-empirical ontologies we do not today accept. In his terms, this is a consideration of the realist claim that "If a theory is explanatorily successful, then it is probably approximately true."[95] With the ultimate aim of supporting the thesis that realism has not in fact been a dominant force in science, he cites

---

[92] Laudan (1984a), p. 106; next quote, p. 104. Note that it is not being claimed here that no one holds views similar to those Laudan is criticising. But those 'realists' to whom he refers are all his contemporaries, and he does not question their logical empiricist presuppositions any more than they do themselves.

[93] Laudan (1984b), p. 219.

[94] Ibid., p. 224; next quote, p. 227.

[95] Ibid., p. 228.

such examples as the crystalline-sphere theories of ancient and me-
dieval astronomy, the phlogiston theory, and the theory of the electromagnetic ether. Laudan's argument here consists essentially in
pointing out the fallacy of asserting the consequent (A implies B;
B; therefore A),[96] coupled with his historical claim that such theories
outnumber theories whose ontologies we today accept by about six
to one.

First, Laudan does not distinguish between those of the theories he
cites which were put forward with the intention that their theoretical
entities be considered actually to exist, and those which were not.
Eudoxus, for example, warned against taking the crystalline spheres
of his astronomical theory as actually existing. Second, it should
be pointed out that such trans-empirical theories as Laudan refers
to would never even have been proposed had their propounders not
adopted a realist attitude. Third, so long as the theoretical entities to
which the realist's theory refers are inaccessible, he should recognise
as well as anyone that the theory is merely hypothetical, that is, that
it is a theory and not a fact. Why should he be expected to claim
it to be correct as a whole (to be true or approximately true), when
he sees it in this light?[97] This idea of *establishing* the correctness
of theories, expressed in terms of determining them to be *true*, is
rather that of the logical empiricists, who have this aim with regard
to theories conceived as universal statements (which has led to their
problem of confirmation). Fourth, it could be argued that each of
the theories Laudan mentions has constituted a step on the way to
what have generally been considered scientific advances of either
an empirical or theoretical nature. Fifth, the theories Laudan refers
to are today generally considered to be incorrect with regard to the
ontologies they suggest, not because this (realist) way of thinking is
itself mistaken, but on the contrary, because this way of thinking has
led to a revision of judgement. It is largely because technological
advance has shown those particular ontologies not to exist and/or
because theoretical advance has led to more realistic ontologies, that

----

[96] What is missed by Laudan here (but also by Galileo as quoted above), as will
be treated in detail in the following chapters, is that there are fundamental constraints on theorising other than simply that a theory produce correct empirical
results.
[97] Cf. above, p. 40, n. 91.

the ontologies of the theories Laudan mentions have been discarded. And sixth, as a matter of fact many theories of this type, again through technological advance, have been shown to be essentially correct.

In sum, what is at issue is the very idea of conceiving of the empirical as having a causal basis in the trans-empirical. In the face of Laudan's argument a realist might well respond that the fact that theorising of a realist type has been done on such a scale as suggested by Laudan's examples supports the view that realism *as a matter of fact* has had an important place in science; that such theorising has paved the way to scientific advance supports the view that realism *should* have an important place in science; and that technological advance has actually shown many such theories to be essentially correct should *secure* realism's place in science.

As is becoming clear, it is unfortunately the case that the modern debate concerning realism and empiricism is largely being carried on within the context of the basic presuppositions of logical empiricism,[98] with realism being misleadingly described as essentially the view that scientific theories are true or approximately true.[99] A variation on this confusion is also manifest in the work of **Nancy Cartwright**, who believes realism in physics to entail what she terms the 'facticity' view of *laws*, according to which all laws of physics are true. She says: "The view of laws with which I begin – 'Laws of nature describe facts about reality' – is a pedestrian view that, I imagine, any scientific realist will hold."[100] In further considering what she takes to be the realist view she says: "A long tradition dis-

---

[98] As suggested by McMullin: "[E]mpiricism is one of the most commodious of all philosophic mansions; so large is it, indeed, that few ever find their way outside it." (1974, p. 129).

[99] This is also the basic view of William Newton-Smith, who believes that " 'Realism' has been used to cover a multitude of positions in the philosophy of science, all of which, however, involve the assumption that scientific propositions are true or false where truth is understood in terms of a cleaned-up version of the correspondence theory of truth." (1981, pp. 28–29). This approach, together with an emphasis on the notion of *sentences*, may also be found in Newton-Smith's more recent work: "Some [scientific realists] state that the sentences of science, be they theoretical or observational, are true or false in virtue of a correspondence or lack of it between what the sentence says and how the world is." (1989, p. 31).

[100] Cartwright (1983), p. 55.

tinguishes fundamental from phenomenological laws, and favours the fundamental. Fundamental laws are true in themselves; phenomenological laws hold only on account of more fundamental ones. This view embodies an extreme realism about the fundamental laws of basic explanatory theories."[101] And with regard to her main work she states: "The primary aim of this book is to argue against the facticity of fundamental laws."

Nowhere, however, does Cartwright identify even one individual who is to belong to the tradition she mentions. The only person the present author can think of who might possibly be ascribed such a view is Whewell, but his name does not appear in the index of Cartwright's work, and her discussion does not meet the issues Whewell raises. Unlike van Fraassen and Laudan, who are arguing against real opponents, it would appear that Cartwright is fighting a straw man. But her straw man, like their opponents, looks more like an empiricist than a realist.

In fact the whole of Cartwright's critical discussion is based on a fundamentally logical empiricist conception of science. She focuses on and accepts without question the notion of truth; she similarly accepts explanation to be paradigmatically performed by laws, not theories; and she does not at all question whether there is an important epistemological distinction between explanation and description, or between explanation and prediction.

Thus, with regard to causality for example, she assumes causal explanation to be "explanation by causal law,"[102] (where causal laws are those which "have the word 'cause' – or some causal surrogate – right in them"). She distinguishes such explanation from theoretical explanation, and speaks of "the tension between causal explanation

---

[101] Ibid., p. 100; cf. also p. 65; next quote, p. 152. It may be mentioned that Cartwright's 'fundamental laws,' in that they are to hold between theoretical entities, bear a certain affinity to scientific causal principles, to be treated in the sequel.

[102] Ibid., p. 29; next quote p. 21. In this general regard cf. also McMullin's critical comments in his (1987), p. 69: "[S]he makes a number of questionable assumptions about the nature of explanation in science. . . . She assumes, for example, that the basic explanatory function in science is carried by laws, not by theories, and that the criterion of explanation is what she calls 'organizing power.'. . ."

and theoretical explanation,"[103] where causal explanations "have truth built into them."

In her more positive contribution however, which she terms the 'simulacrum' account, she begins moving over to a view of science very similar to Harré's, and at the same time revising her conception of causality. She says: "On the simulacrum account, to explain a phenomenon is to construct a model which fits the phenomenon into a theory. The fundamental laws of the theory are true of the objects in the model, and they are used to derive a specific account of how these objects behave."[104] Furthermore, "The simulacrum account is not a formal account. It says that we lay out a model, and within the model we 'derive' various laws which match more or less well with bits of phenomenological behaviour."

And with regard to causality she says: "One important thing we sometimes want to do is to lay out the causal processes which bring the phenomena about, and for this purpose it is best to use a model that treats the causally relevant factors as realistically as possible. ... But this may well preclude treating other factors realistically."[105] And: "All sorts of unobservable things are at work in the world, and even if we want to predict only observable outcomes, we will still have to look to their unobservable causes to get the right answers." *This* is much closer to realism than what Cartwright takes realism to be, viz. the view that 'fundamental' laws must be true. But Cartwright is not aware that her view here is not novel, and towards the end of her book she still thinks that "The emphasis on getting the

---

[103] Ibid., p. 12; next quote p. 91.

[104] Ibid., p. 17; next quote p. 161.

[105] Ibid., p. 152; next quote p. 160. It should be noted that realist model building requires abstraction and idealisation, even with regard to causally relevant factors. Cf. Harré (1970b), pp. 41–42 for his treatment of these notions. There he says, for example: "An abstraction has properties ... fewer than its source-subject. Which properties are chosen depends largely on the purposes for which the model is created. This is because the properties which are not modelled are those which are irrelevant, and relevance and irrelevance are relative to purposes." (p. 42).

Cartwright frequently makes reference, either implicitly or explicitly, to idealisation in her criticism of what she takes to be realism. But see Nowak (1980) for an anti-*empiricist* philosophy of science based on idealisation. This topic is treated in Chapter 5 of the present study.

causal story right is new for philosophers of science; and our old
theories of explanation are not well-adapted to the job."[106]

Another contemporary philosopher, **Ian Hacking**, distinguishes
between what he sees as being two forms of realism, one which he
calls realism about entities and the other realism about theories.

Realism about theories says that we try to form true theories about the world,
about the inner constitution of matter and about the outer reaches of space.
... Realism about entities – and I include processes, states, waves, currents,
interactions, fields, black holes, and the like among entities – asserts the
existence of at least some of the entities that are the stock in trade of physics.[107]

It is Hacking's intention to defend this latter sort of realism, and the
main way he does this is by suggesting that such things as electrons
must exist since as a matter of fact they are manipulated in laboratory
situations. "Electrons are no longer ways of organizing our thoughts
or saving the phenomena that have been observed. They are now ways
of creating phenomena in some other domain of nature. Electrons are
tools."[108]

But as an argument concerning the existence of electrons this just
begs the question. Experimenters do not manipulate electrons – and if
we admitted that they did we would certainly also have to admit that
electrons exist – what they manipulate are pieces of experimental
apparatus. They *assume* such manipulations to have an effect on a
trans-empirical reality of which electrons are a part – and so they
are themselves realists; but nothing in Hacking's argument supports a
stronger conclusion than this. This conclusion *is* noteworthy though
as regards the extent to which actual scientific practice is realist in its
orientation.

This question-begging carries over to Hacking's distinction between what he sees as being two sorts of realism, for he believes
that what he calls realism about entities is justified mainly by the
fact that such entities are actually manipulated by experimenters. But
since as a matter of fact such entities as electrons are not actually

---

[106] Cartwright (1983), p. 162.
[107] Hacking (1984), p. 155.
[108] Ibid., p. 156; also in Hacking (1983), p. 263.

manipulated, his distinction falls in the wrong place for his own purposes. A distinction which *might* be made in this context would be
between realism about theoretical entities,[109] such as electrons, and
realism about what might be termed empirical entities, such as laboratory instruments. A pure phenomenalist such as Mach should deny
the independent existence of both sorts of entity, while a naive realist such as van Fraassen might deny the former while accepting the
latter.

While Hacking's views are not so imbued with logical empiricism
as are those of most of his contemporaries, he nevertheless conceives
of what he calls realism about theories in empiricist terms (that we
try to form true theories), and quite misses the point that the very formulation of theories is a realist occupation whose aim, as mentioned
above, is not primarily to describe the trans-empirical but to explain
the empirical.

To conclude this survey we shall take a brief look at the thoughts
of **Hilary Putnam** on the issue. In a recent contribution, with the
title 'What is Realism?,' he expresses a view which is steeped in
the logical empiricist tradition. An immediate indication of this is
the recurrence of such phrases as "true observation sentences," and
"observational vocabulary," as well as e.g. his suggestion that "we
formalize empirical science or some part of empirical science – that is,
we formulate it in a formalized language L, with suitable logical rules
and axioms, and with empirical postulates appropriate to the body of
the theory we are formalizing."[110] Accordingly, his conception of
realism is also thoroughly empiricist.

With regard to realism, Putnam states: "Whatever else realists say,
they typically say that they believe in a 'correspondence theory of
truth.'" The correspondence theory of truth to which Putnam is referring is based on the ideal-language, referential-theory-of-meaning
idea that all terms occurring in true sentences should have counterparts in reality. On this view, realists claim this correspondence to

---

[109] Here one should keep in mind a further distinction, namely between theoretical
entities and hypothetical entities. While theoretical entities are normally hypothetical
when first introduced, they need not remain so – electrons perhaps being a case in
point.
[110] Putnam (1984), p. 140; next quote, p. 148.

obtain even when the sentence in question cannot be verified or falsified, while empiricists limit meaningful sentences to those whose
correspondence can be empirically checked.

What we see here, however, is not that empiricists and realists are
being distinguished with regard to *correspondence*, but with regard
to verificationism or falsificationism on the one hand, and some transcendent view on the other. In other words, on Putnam's own way of
thinking, both empiricists and realists can hold to a correspondence
theory. What distinguishes them is whether they are willing to think
in terms of a trans-empirical reality.

In the next chapter the first move will be made in the laying out
of an alternative conception of science against the background of
which the empiricism/realism issue can be resolved. It will consist in
the depiction of the core of modern science as consisting in certain
particular principles.

CHAPTER 2

FUNDAMENTAL AND REFINED PRINCIPLES:
THE CORE OF MODERN SCIENCE

The debate over empiricism and realism concerns the very nature of modern science: what it is, or what it ought to be. Empiricism, in its extreme form, claims that there is no reality behind appearances, and that it is the task of science to determine what the appearances are and what the formal relations are that obtain among them. In its weaker form, in which the existence of a trans-empirical reality is admitted, empiricism claims that due to the transcendent nature of this reality no knowledge can be had of it, and it suggests essentially the same methodology as does the stronger form.

Some proponents of both the weaker and stronger versions of empiricism have suggested however that theorising regarding the nature of trans-empirical reality is acceptable to the extent that it aids in extending empirical knowledge, but that such theorising is not to be valued *per se*. On the empiricist view the aim of science is to attain the truth, and the inaccessibility of the trans-empirical, even if it should exist, makes it impossible to establish even probable truths regarding its nature.

For the realist, on the other hand, the attainment of truth is not the sole nor necessarily the ultimate aim of science, it being equally as important to attempt to *understand* why the empirical truths of science are as they are. And such an understanding consists in the acquiring of a conception of how the empirical world comes to be as it is as a result of the nature and workings of the real world. Thus for the realist theorising is an integral part of science, not merely because it may lead to the discovery of empirical regularities, but because it is capable of *explaining* known regularities in terms of a reality lying behind them.

This is the point at which the issue is usually left, empiricists taking the stance that understanding is not understanding unless it is based on knowledge, and realists retaliating that it is debatable whether what the empiricists take to be factual knowledge does not itself involve the assumption of certain unknowns.

In the present work, however, a novel approach will be taken to the issue in the attempt to provide deeper insight into its nature. In what follows, the conceptual underpinnings of empiricism and realism will be considered against the background of certain potentially basic metaphysical presuppositions of modern science. These presuppositions will have application to both the empirical and theoretical aspects of science, not only indicating the basis of the difference between empiricists and realists regarding theorising, but also revealing a fundamental divergence in their ways of conceiving of research on the empirical level.

Now of course there are views, though perhaps not so popular today, according to which science is a presuppositionless activity. Our argument for a particular version of the contrary view will consist essentially in showing its explanatory power when the fundamental presuppositions of science are specified and their relation to science indicated.[1] This will consist partly in showing how the present view can contribute to understanding the empiricism/realism issue, and partly in clarifying other aspects of science as well.

What are the central candidates for being the fundamental presuppositions of modern science? How should they be conceived? Or, more particularly, should they be taken as concerning how to go about doing science, or as concerning the nature of the subject-matter of science?

This itself is a major issue – one not unrelated to that of empiricism vs. realism. It has commonly been thought that modern science is to be distinguished by its method – in particular, by the experimental method – which would imply that its fundamental presuppositions are methodological. This is not the approach taken here, however. Rather, the basic presuppositions – or *principles* – of mod-

-----

[1] Hübner expresses our version of the contrary view thus: "To the degree that there is cognition by means of scientific concepts, this cognition will always depend on metaphysical presuppositions. Thus metaphysics is by no means dead as has often been announced; on the contrary, metaphysics is, as we see, even unavoidable." (1988, pp. 116–117).

ern science are here seen to be ontological, i.e. to concern the nature of the subject-matter of science. The reason for this can perhaps to some extent be made clear by assuming the contrary, and taking e.g. the presupposition of the epistemological viability of the experimental method as being central to modern science. But then the question may be asked as to why this method should be considered to give epistemologically valuable results. Why should these results be of more than momentary significance? Why should they be replicable? The answer to these questions, as will be presented in detail in the next chapter, is that it is assumed by scientists that nature is, in a particular sense, uniform. Thus, in this case, the adoption of the experimental method itself involves ontological presuppositions. More generally, we should say that it is only when one has some conception of the nature of the subject-matter being investigated that it makes sense to commit oneself to a particular methodology for investigating it.[2] And, while one cannot use the nature of the methodology to explain the assumed nature of the subject-matter, as will be demonstrated in what follows, reference to certain general assumptions regarding the subject-matter can explain the adoption of a particular methodology.

So the basic presuppositions or principles of modern science will here be taken to be ontological in nature, that is, to be very general assumptions regarding the nature of what modern science sets itself to investigate. More particularly, they will be taken to be three in number, and to be the specific ones presented below. Before going on to consider them, however, two points must be emphasised. The first is that these principles will be taken in this work as constituting, in their relatively pure form, the *core*, rather than, say, the *foundation*, of modern science. Thus they are to be conceived as together delineating an ontological paradigm or ideal – an ideal conception of reality (whether empirical or transcendent) which may generally be presupposed in the doing of science, but which need not be accepted without qualification in all activities deserving to be called scien-

---

[2] In this general regard, cf. Crompton (1992), pp. 149–150: "[S]uch a conception of the world [cannot] begin to be constructed in the absence of a general idea of what it will be like when it is finished (if it is ever to be finished), an idea which will thus be a conception of what reality is generally like: a metaphysics."

tific.[3] This way of thinking is akin to Wittgenstein's as it relates to the function of paradigms in linguistic behaviour. Thus it is here not a yes-or-no matter of either accepting or rejecting a particular principle or set of principles in doing one's work, but rather of *tending* to accept certain principles instead of others, and of conceiving of what one is doing in terms of the principles one tends to accept.

The second point is that what is here meant by modern science is science since the time of Galileo and the Scientific Revolution, as may be contrasted, for example, with Platonic or Aristotelian science, or with psychoanalysis conceived as a science. Though modern science has of course gone through many changes since the time of Galileo, we still think of it as essentially the same activity now as then, which is distinct from other real or imaginable activities by which one might attempt to obtain knowledge or understanding of reality. Thus modern science is here seen as being but one way of going about investigating the world. In this regard the present work differs from most other efforts of a similar nature, for in them it is seldom the case that any distinction is drawn even implicitly between modern science and other potential forms of science, and it is almost invariably presupposed further that (modern) science constitutes the best way to go about investigating reality, the problem being one of saying what in particular this best way actually or ideally consists in. On the other hand, no position is being taken at this stage as to the ultimate viability of modern science as an epistemological activity; nor is any prescriptive stance being assumed with regard to the empiricism/realism debate, though it will be argued that one of these two views is the more in keeping with the essential nature of modern science. Rather, the main aim of the present work is to capture the essence of modern science, and in so doing indicate the positions of empiricism and realism with regard to it.

What are to be presented below then are three candidates for being the core presuppositions of modern science.

---

[3] This paradigm or ideal is what Toulmin refers to as an 'ideal of natural order' (1961, p. 38; see also e.g. pp. 56–57 and 81).

1. THREE PRINCIPLES CENTRAL TO MODERN SCIENCE

A. The Principle of the Uniformity of Nature

It has generally been recognised by those who have devoted themselves to considering such matters that the principle of the uniformity
of nature is of central concern to modern science. Here we shall attempt to make this principle explicit in the paradigmatic guise it
adopts for modern science, and indicate some of its implications.

The principle of the uniformity of nature concerns change, and is
usually understood to mean that natural change is lawful, or takes
place according to rules. It thus implies a deterministic conception
of change, though this determinism need not be strict. For example,
the rules according to which change takes place might on occasion
be broken, while the principle still retain its basic validity – such
a breaking of the rules perhaps constituting a *miracle*; or it may be
that the principle apply only to broad categories of change, setting
deterministic limits within which relatively undetermined change can
take place – as we assume when we grant ourselves and fellow humans
free will; or the laws according to which change takes place may
regulate the behaviour of ensembles but not their constituents taken
individually, as is generally thought to be the case with regard to
probabilistic laws.

In some form or other the principle of the uniformity of nature
has been assumed by human beings of all cultures and all historical
epochs in their daily lives. Without the adoption of this principle,
and an assumed awareness of some of the rules according to which
natural change takes place, there would be no basis for reasoned
action concerning the future, whether near or distant. As a basic
presupposition of modern science, however, the principle takes on a
particular form which has specific implications regarding the nature
of the world which science investigates.

The form of the uniformity principle which will interest us in what
follows is that according to which similar states of nature are followed by similar states. Thus if a particular state A is similar to a
state B, and A is succeeded by A', then B will be succeeded by B',
where A' and B' are similar. Taken to an extreme in this regard, the
principle suggests that exactly similar states are followed by exactly

similar states, which expresses implicitly a form of strict determinism with regard to change in nature.

Whether one takes the principle in a stronger or weaker sense, it is of course important to establish what should be the case in order for two states to be or not to be similar. A partial specification in this direction, which is of particular relevance to modern science, is that location in space and time are not constituents of a state. In other words, whether or not states *A* and *B* are similar is in no way dependent on their positions in space or time *per se*.[4] This has important implications for how *space* and *time* are conceived, suggesting that both are in a sense homogeneous.[5] Thus, for example, the Aristotelian conception of space, according to which the universe has a centre towards which heavy things tend, is excluded.

The idea of homogeneous space and that of strict determinism were generally accepted in the classical period of modern science, or more particularly, modern physics. In the twentieth century however both of these conceptions have come to be questioned. On at least one interpretation, the general theory of relativity implies a denial of the homogeneity of space; and many people believe that the results of quantum mechanics rule out a strict determinism in nature. One of the theses of the present work, as will be returned to below, is that such changes can be well understood in terms of changes in the accepted form of the core principles of modern science, in the present case, the principle of the uniformity of nature.

The principle of the uniformity of nature is to apply to change, which gives rise to the question of what change itself is.

------

[4] E. P. Wigner expresses this specification thus: "[G]iven the same essential initial conditions, the result will be the same no matter where and when we realise these;" or, "in the language of initial conditions, as the statement that the absolute position and the absolute time are never essential initial conditions." (1967, p. 4).

[5] This conception, and the form of the uniformity principle it presupposes, can be traced at least as far back as to early Greek atomism, with its conception of eternal time, infinite space and predetermined change. In this regard see e.g. Čapek (1961), p. 123, and Bailey (1928), pp. 120–121.

B. The Principle of Substance

There are basically two conceptions of change, one according to which everything is in flux and nothing is permanent – a conception associated with Heraclitus – and the other according to which all change is but change in a perpetually existing substratum or substance – a notion that can be traced as far back as to Thales. A compromise between the extremes of there being either no substance or a perpetual substance is the view according to which substances (independently existing realities which are not the properties of anything else) can exist for finite periods of time – as may be found e.g. in Aristotle's notion of a primary substance.

The conception that is of particular relevance to modern science is that according to which substance exists perpetually and change is but an alteration of the substance. An important corollary of this conception is that no (portion of) substance either comes into or goes out of existence. This conception might be termed the principle of the perpetuity of substance, but for ease of reference we shall simply call it the principle of substance, bearing in mind the notion of substance intended.[6]

It is possible to interpret the principle of substance in different ways. The substance referred to may be considered to be material, or be taken as being immaterial or formal,[7] or as being, for example, God.[8] As regards modern science however, substance is classically

---

[6] This conception recurs throughout Presocratic thought – e.g. in the works of Anaximander, Anaxagoras, Empedocles, Melissus and Diogenes of Apollonia – receiving its definitive ancient expression in Greek atomism. For a relatively modern expression of the principle, see e.g. Kant (1781, 1787), A 182, B 224: "In all change of appearances substance is permanent; its quantum in nature is neither increased nor diminished." One can question however whether Kant is in the position to adopt such a principle given the idealistic nature of his philosophy – a point to be returned to in the following chapter.

[7] Plato conceives of substance (essence) as formal and eternal; see e.g. *Republic*, 7.527a–b, *Timaeus*, 51e–52a. It is perhaps worth noting in the present context the difficulty of employing such a notion in accounting for change.

[8] Spinoza takes God to be eternally existing substance: cf. his (1678), esp. prop. XVI and XIX.

conceived to be material,[9] and change to consist in the *motion* of (part of) the substance. This classical conception can be refined in one of two ways, such that substance is conceived as occupying either all or only part of space, i.e. as constituting a plenum or as not. One further refinement suggests that the whole of one place cannot be occupied by more than one portion of substance at a time, and that all of a portion of substance cannot occupy more than one place at a time (cannot be in two or more places at the same time). Also, on this conception motion is itself considered to exist perpetually as well (as are time and space, which it implies), so to that extent it may also be considered a substance.[10]

Note that we are here speaking in terms of a particular paradigm – viz. that of classical modern science – and that in the history of science the notion of substance has undergone evolution: for example, the notion of a field has developed whereby two fields, both being physical existents, can occupy the same space. The principle itself however, according to which all change consists in the alteration of the underlying substance, has remained the same. Thus we see that the specific way in which substance is conceived can evolve as regards a particular paradigmatic conception, while the principle retain its applicability.

Similarly, the way in which substance is conceived in one scientific discipline might well differ from the way it is conceived in another, while the principle of substance be applicable to both. In fact, it could be argued that the substances of different disciplines must differ in some significant respect in order that the disciplines actually be different. Furthermore, the substance of a discipline can take different forms.

What we find in the case of modern science is that the substances of certain disciplines are presupposed by those of others, placing the disciplines in a hierarchy of ontological dependence. Thus the sub-

---

[9] See e.g. Bacon, where he expresses the substance principle thus: "'nothing is made from nothing, nor can anything be reduced to nothing'; but the actual quantity of matter, its sum total, remains constant, being neither increased nor diminished." (1620, Book II, Aphorism 40).

[10] The idea of the substantiality of motion is discussed in some detail in Čapek (1961), pp. 71–77.

stance of chemistry presupposes that of physics, that of biology presupposes that of chemistry, and that of the social sciences presupposes
that of biology.

Another point regarding the flexibility of the principle of substance
is that it can be relativised to a discipline in the sense that the substance
of a discipline need only exist perpetually from the point of view of
the discipline itself. Thus, for example, if we take the substance of
sociology to be society, we do not mean that society exists perpetually,
but that there being a society is a prerequisite for the pursuit of the
discipline, so that there must always be a society as far as sociology
is concerned.

Assuming that change consists in the uniform transformation of
substance, it may be asked how that transformation is brought about.

C. The Principle of Causality

The principle of causality states that change is caused. Like the
principle of the uniformity of nature, it has been and is held in some
form or other by human beings of all cultures. The nature of the causes
thought to operate can differ greatly however. Perhaps the most drastic
divergence in this regard – and one which is of paramount importance
for modern science – is between conceiving of a cause either as being
what we would today term natural, or as being supernatural. Thus, for
example, while all would agree that earthquakes are caused, ancient
Egyptians and Mesopotamians, and people of primitive cultures today, might well claim that the cause of a particular earthquake was
an irate god. Thales and certain other Presocratic philosophers, on
the other hand, as well as scientists and educated laypeople of today,
would claim that the earthquake in question resulted from specific
natural, or more particularly physical causes.

The idea of supernatural causes has no place in modern science
(though the attempt has been made to include them, perhaps most
notably by Newton). But all 'natural' causes need not be conceived
of as being physical, and there are important alternatives to be considered, the foremost of which is the idea of a *formal* cause. Much
of today's struggle between realism and empiricism is a struggle over

the question whether the physical or the formal notion of cause should be given primacy. Concerning the early historical backgrounds of the two notions, it may be said that the physical notion originated with Thales and was further developed largely by the Ionian philosophers, while the formal notion originated with Pythagoras, became the cornerstone of an extremely influential philosophy in Plato, and adopted a particularly noteworthy guise in Aristotle's notion of a final cause.

One line that has been adopted by those advocating the fundamental status of formal causes of a particular sort is in effect to identify the principle of causality with the principle of the uniformity of nature, stating that causality is essentially nothing other than constant conjunction (cf. e.g. Mill, as treated in Chapter 1). From a broader perspective however, in which one is attempting to understand both the empiricist and realist approaches to science, it is important to distinguish the two principles. Their essential difference lies in the fact that the principle of causality does not concern lawfulness, or the following of a rule, but cause, or the production of an effect. Where the principle of the uniformity of nature implies a regularity or determinism in natural change without indicating the reason for that change, the present principle indicates the reason without implying the regularity. Thus, for example, the principle of causality is quite in keeping with the free will of that to which it is applied, while it is at least arguable that the uniformity principle is not,[11] and miracles can be conceived of as being caused (by God), in spite of their violating the uniformity principle.

However, the two principles do fit well together (so well, in fact, that they are seldom distinguished),[12] with the causality principle in a way providing a content for which the uniformity principle provides

[11] In this regard, see Born (1951, p. 8): "I think one should not identify causality and determinism. The latter refers to rules which allow one to predict from the knowledge of an event A the occurrence of an event B (and vice versa), but without the idea that there is a physical timeless (and spaceless) link between all things of the kind A and all things of the kind B."

[12] Cf. e.g. Hempel (1962), pp. 104–105: "When an individual event, say b, is said to have been caused by a certain antecedent event or configuration of events, a, then surely the claim is intended that whenever 'the same cause' is realised, 'the same effect' will recur."

a form. Unified in a particular way, these two principles state that all change is caused in a regular fashion, i.e. that causes are regular in their effects,[13] where position in space and time are not causes.[14]

Conjoining the principle of substance to these two principles, we obtain the idea that causality acts regularly through the action of one portion of substance upon another, and that change consists in the relocation of substance. If, as on the paradigmatic conception expressed earlier, motion is taken to be perpetual, then it too does not come into or go out of existence, but changes only with regard to its form. In this case then causes act so as to change the form or *state* of the motion of (a part of the) substance.

On the classical conception the action of one portion of substance upon another consists in the two colliding. Thus on this conception cohesion is to be understood in terms of external bombardment, and any form of action at a distance is excluded. This conception has faced great difficulties of application, beginning as early as in the case of Newton's theory of gravitation. Neither Newton nor his successors have taken easily to the idea of action at a distance however,[15] and while the collision requirement in the case of causal relations has been set aside, the idea of contact has not. Thus throughout the history of modern science there has been a felt need on the part of scientists to determine how events take place as a result of the operation of causes which are adjacent to their effects. This may be expressed in terms of the *contiguity principle* of causality, according to which causes are to be contiguous to their effects, both temporally

---

[13] This has been expressed by Herman Weyl as: "[S]ufficiently like causes lead to nearly like effects." (1949, p. 189). Cf. also Aristotle, *On Generation and Corruption*, 336ᵃ25–30: "[N]ature by the same cause, provided it remain in the same condition, always produces the same effect."

[14] Maxwell expresses this specification thus: "The difference between one event and another does not depend on the mere difference of the times or the places at which they occur, but only on the differences in the nature, configuration, or motion of the bodies concerned." (1877, p. 13). Aristotle, on the other hand, maintains the causal efficacy of place: "Further, the locomotions of the elementary natural bodies – namely, fire, earth, and the like – show not only that place is something, but also that it exerts a certain influence." (*Physics*, 208ᵇ9–11).

[15] Regarding Newton, see p. 102, n. 8, below.

and spatially.[16] A weakened form of the contiguity principle is the *proximity principle*, which states that the nearer the cause (normally, in space), the greater the effect.

Other refinements of the general principle of causality of direct consequence to modern science are that the size of an effect be proportional to that of its cause, and that the action of every cause be balanced by an equal and opposite reaction.[17]

As mentioned earlier, it is important to keep in mind that the above principles are here intended to be conceived as constituting the *core* of *modern* science. Thus they are not to be thought of simply as general statements about the nature of reality – statements the truth of which may or may not be presupposed when one is engaged in epistemological activity generally. Rather, in the particular form emphasised above, they are more to be thought of as constituting a conceptual paradigm – a paradigm which determines a particular kind of epistemological activity. But this is not to say that one cannot be led to question them, either on the basis of general philosophical considerations or on the basis of pursuing the activity for which they constitute the ideal. Thus, for example, these particular principles can be criticised for difficulties had in applying them to psychological phenomena, conceived quite generally; and they can be criticised for their apparent incompatibility with certain physical phenomena, which have been discovered by actually following them.[18] They can also be criticised independently of their applications, along lines having to do with their inherent conceivability. But

---

[16] Concerning this refinement, see again Born (1951, p. 8): "[I]t is generally regarded as repugnant to assume a thing to cause an effect at a place where it is not present, or to which it cannot be linked by other things; I shall call this the principle of contiguity." The same principle is referred to by Weyl as the principle of nearby action (1949, p. 189), and by Harré as the principle of the proximity of some cause to an effect (1970b, p. 73). In this regard Enriques speaks of a "fundamental exigency of our spirit . . . to try to represent actions (the play of causes) as propagating by contiguity in space and time." (1934, p. 24).

[17] For a detailed discussion of these two refinements, and their relation to the principle of causality, see Whewell (1847), Part 1, pp. 177–185.

[18] As expressed by Enriques: "the evidence of principles does not in any case constitute an *a priori* proof that shall be valid in the face of possible contrary experiences." (1906, p. 246).

in this regard it must be pointed out that in neither the natural nor the social sciences do we at present have alternatives of equal simplicity, coherence and generality to take their place, and that if we did, we might then ask whether we still had to do with what we today call 'modern science.'[19]

Thus, just as important as the question of the extent to which these principles have been accepted or rejected through the history of modern science, is the idea that they provide the fundamental *categories* of scientific thought.[20] In this way they constitute the core of the conceptual framework in the context of which science is pursued, so that even though they may have been followed in a less qualified form more closely in classical than in contemporary modern science, what *is* followed in contemporary science is nevertheless framed in terms of them.

In a similar vein, the reader might be reminded here that it is by no means the purpose of the present work to argue either for the general viability of these principles, or for the excellence of modern science. The present interest is rather in attempting to explain the nature of science, leaving aside questions as to its value, epistemological or otherwise, until our explanation is complete.

---

[19] It may be of interest to cite G. H. von Wright in this context. He says: "The reader has perhaps fastened on my rather restrictive use of the word 'science.' By the scientific conception of the world I mean more particularly the worldview of the *natural* sciences. I do not want to consider the conceptions of man and his society given by the so-called human sciences as parts of 'the scientific worldview.' There simply does not exist any comprehensive conception in the sciences of man with anything near the same unity and claim to generality as in the natural sciences." (1986, p. 11). Part of the task of the present work is to show why this is so.

[20] The importance of categories with regard to scientific thought is also of concern to Einstein where, with reference to realism, he says: "We are here concerned with 'categories' or schemes of thought the selection of which is, in principle, entirely open to us, and whose qualification can only be judged by the degree to which its use contributes to making the totality of the contents of consciousness 'intelligible.'" (1949a, p. 673).

2. REFINEMENTS OF THE PRINCIPLES IN SCIENCE

The three ontological principles presented above are extremely general, so general in fact that they are seldom, if ever, explicitly referred to in the doing of science. All of them have been refined, however, and in certain cases have taken the form of explicit principles – a form determined in part by the nature of the subject-matter being investigated, and in part by the stage of the investigation.[21] It should be pointed out however that such refinement does not consist in the refined principles' being formally *deduced* from the more fundamental principles, but rather in their being *drawn* from them in such a way as to allow the fundamental principles to be *applied* in particular cases, a process which may involve their *qualification.* Furthermore, as will be seen in what follows, refined principles are in many cases the result of the application of more than just one of the fundamental principles, and for this reason involve an interweaving of their various concepts.

A. Principles of Spatial and Temporal Invariance

The clearest explicit manifestations of the principle of the uniformity of nature in modern science are principles of spatial and temporal invariance (and covariance). Understood as being about space and time, the principle of the uniformity of nature is itself the most general scientific principle of this kind.[22]

Within physics, the principle becomes refined and qualified in various ways. Thus we have the invariance or covariance (invariance of form) of particular physical laws with respect to different states of motion or choice of coordinate system, for example the covariance principles of special and general relativity.

Another scientific principle, which is seldom explicit but is nevertheless widely assumed in science, follows directly from the general

---

[21] As regards, e.g., the principle of causality, Whewell takes up the role of experience in its refinement (1847, Part 1, pp. 166–167, 245–254), and the different forms it can take in different sciences (ibid., Part 2, pp. 97–103).

[22] According to Wigner, this principle "is the first and perhaps the most important theorem [*sic*] of invariance in physics. If it were not for it, it might have been impossible for us to discover laws of nature." (1967, p. 4).

principle of the uniformity of nature. It is the principle of the indifference of spatial orientation, understood to mean that similar states are followed by similar states independently of their orientation in space. Taken as being about the nature of space, this principle has the implication that space is isotropic.

A similar principle following directly from that of uniformity is the principle of the relativity of magnitude,[23] or the principle of scalar indifference, which is to the effect that states similar in every respect but size are succeeded by proportionally similar states. In terms of space it implies not only that space is homogeneous, but that any volume of space is like any other regardless of size, and that there is no smallest or largest volume of space (space is both inwardly and outwardly infinite) – a conception some would consider repudiated in twentieth-century physics.

B. Conservation Principles

The principle of substance is manifest in modern science in the form of various conservation principles.[24] Perhaps the most straightforward example is the fundamental principle of chemistry, namely, the principle of the conservation of matter, according to which in chemical processes matter never changes in quantity but only in form. In physics the corresponding principle is that of the conservation of mass, which has been refined to become the principle of the conservation of energy, the notion of energy involving that of mass as well as those of the quasi-substantial entities time and space.[25]

---

[23] This principle is dealt with in detail in Čapek (1961): see pp. 21–28 and 46–47. Tait expresses it thus: "*[T]here is no such thing as absolute size*, there is relative greatness and smallness – nothing more." (1876, p. 284).

[24] In this regard cf. e.g. Hutten (1962), p. 73; cf. also Čapek (1961), p. 137: "Various conservation laws are different variations of the same basic theme: a certain substantial quantum remains constant while its constitutive parts change their places in space. The quantity of matter remains the same, but its parts change their spatial relations; the quantity of motion is preserved, but its spatial distribution changes; the quantity of energy remains constant, but its spatial distribution varies."

[25] We note here that in physics conservation and invariance are closely related notions. Conservation always implies an invariance of something (the quantity of the substance in question); and invariance always implies conservation (of form, if not of substance). As expressed by Enriques (1906, p. 137), "the general idea of sub-

These notions are used in defining those of closed and isolated physical systems, a closed system being one which *matter* neither enters nor leaves, and an isolated system being one which *energy* neither enters nor leaves.

Here we see how the one fundamental principle can concern different aspects of reality (matter vs. mass or energy) when employed with respect to different sciences, and how when employed within one science it can evolve (conservation of mass to conservation of energy) as the aspect it concerns is reconceptualised. In this latter regard it is noteworthy how the notion of substance in physics has come to be intimately associated with the notion of force, such that conservation of energy might also be considered under the heading of principles of dynamics.

C. Principles of Dynamics

The principle of causality, with certain of its refinements mentioned above, is most clearly manifest in science as principles of dynamics, the most fundamental of which is the principle of inertia: that a body will continue in its state of rest or uniform motion in a straight line unless acted upon by a force.[26] Here the notion of force, or 'interaction,' is the physical specification of the more general notion of cause;[27] and change consists in change of state (rather than change of position) due to the presupposition of the substantiality of motion. Thus the principle of inertia implies that there are no uncaused changes of state.

---

stance takes on a concrete form through the recognition of certain invariants." It is important for present purposes however to maintain the commonly recognised distinction between the two notions.

[26] This principle is in essence presupposed by early Greek atomism (as is the principle of the conservation of momentum). In this regard see e.g. Enriques (1934), pp. 58–60, and Čapek (1961), pp. 71–77. Its importance for modern science is emphasised by Dijksterhuis: "[T]he law [principle] of inertia is not just a detail of the new worldpicture, but one of the foundations underlying the most essential parts of the system." (1959, p. 348).

[27] Much attention is devoted to this point in Whewell (1847), Part 1, pp. 164ff., 215ff. In this regard cf. also Weyl (1949), p. 149, Born (1951), p. 12, and Bigelow et al. (1988).

The further refinement of the fundamental principle of causality which states that causes be contiguous to their effects is perhaps most clearly expressed in contemporary physics as the principle of locality, according to which causal influence cannot be propagated at a speed greater than that of light; and the refinements suggesting that effects be proportional to their causes, and that the action of every cause be balanced by an equal and opposite reaction, correspond to the second and third laws of Newtonian mechanics.[28]

As may be seen from the above, Newton's laws or axioms of motion are here being conceived as explicit scientific principles, to be distinguished both from empirical laws and from theories. Thus Newton's mechanics, or similarly classical thermodynamics, is not here to be conceived of as a theory, but as a set of refined principles.

The fundamental principles mentioned above, as well as their counterparts in science, concern four main notions: space, time, substance and causality – the first principle having to do with space and time, the second with substance, and the third with causality. These notions thus obtain a special status in science, and may be seen as being among its most basic categories, in terms of which all scientific thought takes place. From a consideration of the above principles it would appear, however, that even more basic are the notions of constancy and change, and that they perhaps constitute the two most fundamental categories of modern-scientific thought.

3. FOUR WAYS PRINCIPLES FUNCTION WITH RESPECT TO SCIENCE

To this point we have described certain fundamental principles concerning the nature of reality, and have indicated particular explicit principles in terms of which they are manifest in science. We shall now go on to give a more general description of the various roles played by such principles with respect to science, whether it be in

---

[28] The interweaving of the fundamental principles is manifest here in the intimate relation between the principles of dynamics and the principles of the conservation of momentum and energy. In this regard cf. e.g. Tait (1876), pp. 33–38, Čapek (1961), pp. 76 and 138, and Hutten (1956), pp. 134–135.

their fundamental or their refined form. We note that these roles do not exclude one another, and so such questions as to whether principles function in one of the ways treated here rather than another do not arise.[29]

There are at least four different ways in which principles can be seen to function with regard to science. The **first** way is that they determine what is to be conceived as ontologically *necessary* or *possible* within the enterprise or certain of its sub-disciplines. Thus, for example, the principle of the uniformity of nature rules out the possibility of miracles, as these are normally understood. Similarly, the principle of substance implies that it is impossible for substance to arise from nothing or become nothing. In its explicit form in physics as the principle of the conservation of energy it says the same regarding energy. Given the present state of physics, physicists would consider it impossible that energy increase or decrease in an isolated system, and significant steps have been taken to avoid such a conclusion in their reasonings. These range from the suggestion that physical objects increase in mass when they move, to the suggestion of the existence of previously unheard of particles, such as the neutrino.

It is natural in this context to think of a particular cosmological theory, namely the steady-state theory, as affording a counter-example to our thesis. It must be pointed out however that in being a cosmological theory, the steady-state theory really stands outside of present considerations. This is because cosmology itself stands apart from modern science. Though cosmology employs notions taken from physics, by its very nature it involves a probing into the form adopted by basic ontological principles. In the case of the steady-state theory this probing led to the idea that every so often, in intergalactic space, hydrogen atoms appear out of nothing in sufficient quantity that the mean density of the universe remain constant despite the apparent drifting apart of the galaxies. Such speculations

---

[29] This kind of question regarding principles has been considered important by a number of thinkers. Poincaré, for example, has been concerned as to whether or not principles are definitions; and many questions in analytic philosophy concern the status of particular expressions without at all considering that they can have a different status in different contexts.

are the makings of cosmology. But it is of some interest to note how physicists reacted to this theory. Their criticism was not directed at the idea that matter should condense out of nothing, but that in so doing it would apparently contravene the principle of the conservation of energy!

The notion of what is scientifically possible or necessary may be contrasted with that of what is logically possible or necessary. The logical notions are based on the idea that nothing can at the same time both have and not have the same attribute:[30] that this is so is necessary; its contrary is impossible; and anything that does not infringe it is possible. In logic this takes the form of the principle of non-contradiction, which itself presupposes a specific kind of language in order to be applicable. Thus a particular state of affairs is logically possible if its linguistic description does not imply a contradiction, and is logically necessary if its description expresses a tautology. Given that all of the pertinent states of affairs are adequately describable in the relevant kind of language – itself a moot assumption – this makes the realm of what is logically possible much greater than that of what is scientifically possible, and the realm of what is logically necessary much smaller than that of what is scientifically necessary.[31]

Through delimiting what is to be conceived as possible, principles set limits on the scientist's way of thinking, that is, they provide the structure of *scientific rationality*, which is their **second** function.[32] Rationality in general is not equivalent to logic, nor is scientific rationality equivalent to logic.[33] It is principles that determine the point

---

[30] Cf. Aristotle, *Metaphysics*, 1005^{b}19: "[T]he same attribute cannot at the same time belong and not belong to the same subject in the same respect."

[31] Concerning the divorce of science from logic in this context, cf. Whewell (1860), p. 342: "I will not pretend to say that this kind of necessity [logical necessity] is represented by any of those Fundamental Ideas which are the basis of science."

[32] Cf. David Knight's comment with regard to the conservation principles: "They are the kind of assumptions which make whole sciences possible and rational." (1986, p. 162).

[33] A similar point has been made by Toulmin, where he says: "Having confined themselves to questions about the formal relations between the *propositions* of science, the philosophers concerned have had no alternative except to identify 'rationality' with 'logicality.'" (1973, p. 891).

beyond which it no longer makes sense to ask for a reason; and in science this point consists in the indication of how what is to be explained is but a manifestation of the principles at the core of the enterprise.[34]

We might consider as an example the principle of induction as it applies to science. Unlike the principles discussed above, which may be termed ontological in that they concern the nature of reality, the principle of induction is more a principle of reason, in that it concerns the sorts of conclusions that might reasonably be drawn in certain situations. On what is this principle based, if anything? As has been argued by Hume and reasserted by countless others, the principle of induction cannot be based on experience, for in order to call upon experience in such a way as would support it, the principle itself must be presupposed. And it cannot be based on logic, for from a logical point of view it involves an invalid form of inference. So it apparently has no basis. On the present view, on the other hand, as far as modern science is concerned, the principle of induction rests on the principle of the uniformity of nature.[35] The ontological principle that similar states are followed by similar states supports the principle of reason which allows one to conclude that this will be so in a particular case where knowledge of the succeeding state is lacking. Here we see once again how logic misses the mark with regard to science, scientific principles being more liberal than the logical canons as regards what can rationally be inferred.[36]

Thus principles determine what is to be considered scientifically possible and necessary, as well as what is to be considered scientifi-

---

[34] Brian Ellis suggests how this can be so in the case of the principle of inertia: see Ellis (1965), pp. 45–46.

[35] That the principle of induction is independent of that of *causality* is stressed by Born (1951, p. 7): "I have to make it clear why I distinguish this principle of induction from causality. Induction allows one to generalise a number of observations into a general rule: that night follows day and day follows night, or that in spring the trees grow green leaves, are inductions, but they contain no causal relation, no statement of dependence."

[36] The irrelevance of logic to science is also emphasised by Bacon, where he claims that the "[f]aulty demonstrations...we see in dialectic [logic] have the effect of giving over and enslaving the world entirely to human thought, and thought to words." (1620, Book I, Aphorism 69; see also Aphorisms 12, 63 and 82).

cally rational, thereby as much as fulfilling their **third** function, which is to set guidelines for the actual doing of science. In other words, ontological principles determine the methodology of science. They constitute the presuppositions scientists have in their work, presuppositions which give that work direction and focus. This makes it idle simply to advance *methodological* principles concerning how to go about acquiring knowledge or understanding, without giving them an ontological grounding, as has frequently been done throughout the history of reflection on science.[37]

As will be shown in the following two chapters, ontological principles guide research throughout a branch of science, whether that research be empirical or theoretical. On the empirical level, the principle of the uniformity of nature underlies the whole of the experimental method, suggesting to empirical scientists that they attempt to determine *in particular* which similar states are followed by which, i.e., that they attempt to discover *empirical laws*. This involves a controlling of the conditions in which phenomenal changes occur in such a way as can be *replicated*, and in registering those changes objectively. It may not be evident, however, that the changes registered in this manner are in keeping with the ontological principles as a whole, or with certain of their implications or refinements. When this connection is not clear, theoreticians, for their part, attempt to make it so by representing the changes in terms of a *specific* causally efficacious ontology in which the principles are strictly adhered to; and if they succeed, they may be said to have *explained* the changes. But if, on the other hand, every possible way of theoretically reconciling the phenomena with the ontological principles in question appears blocked, the principles may themselves come to be questioned, and eventually be emended or perhaps even replaced.

A **fourth** way in which principles can function with respect to science is as definitions of its basic concepts,[38] thereby delimiting the

---

[37] To mention but two examples, Newton's 'Rules of Reasoning in Philosophy' (1687, pp. 398–400), and Popper's 'Three Requirements for the Growth of Knowledge' (1962, pp. 240–248).

[38] In this regard, cf. Poincaré's famous saying: "Principles are conventions and definitions in disguise." (1902, p. 138).

nature either of the enterprise as a whole, or of its various sub-
disciplines.[39] A change in the principles of a science may thus be
seen as indicating a change in the nature of the science itself; and
a sufficiently drastic change can lead one to question whether it is
still the same science. Thus, for example, the fundamental changes in
physics occasioned by relativity theory and quantum mechanics are
not to be seen as changes of theory, which would not affect the nature
of physics itself, but as changes in the basic principles of physics,
which have the effect of redefining the discipline.

Together with the third way discussed above, this may provide
a profitable means of understanding Thomas Kuhn's notions of
paradigm and scientific revolution.[40] Here the paradigm around which
a science is constructed, and in terms of which its basic notions are
defined, consists of nothing other than its principles, their paradigmatic aspect being manifest in their methodological influence. On
this way of thinking scientific revolution consists in a change in or of
a particular principle or set of principles, the more fundamental the
change, the more thoroughgoing the revolution. Here too we have a
way of understanding the 'meaning change' alluded to in discussions
of incommensurability, such terms as "mass" in Newtonian mechanics being 'redefined' by the principles of relativity theory. And in a
similar vein we can say that a field of inquiry becomes what Kuhn
terms a mature science when its basic principles are agreed upon by
all practitioners, and that 'normal science' consists precisely in the
detailed application of those principles to reality.

Similarly, Paul Feyerabend's 'methodological anarchism' may here
be seen not as the view that people should epistemologically do simply as they please, but that research based on a *variety* of different principles or sets of principles should be undertaken and supported; and it may be contrasted with the unity of science pro-

---

[39] Cf. Kant (1786), Ak. 472–473: "But of the greatest importance for the benefit of
the sciences is severing heterogeneous principles from one another and bringing each
kind into a separate system, so that each may constitute a science of its own kind."
[40] Cf. Kuhn (1962), p. 34, or p. 103: "[T]he reception of a new paradigm often
necessitates a redefinition of the corresponding science."

gramme, which in present terms advocates that all epistemological or scientific activity be based on but one set of principles.

4. ON THE EPISTEMOLOGICAL STATUS OF THE PRINCIPLES OF SCIENCE

The idea that science rests on principles is not itself new, having been expressed in various forms by a number of notable philosophers and scientists in the past. The novelty in the present view consists in the way the principles and their relation to the rest of science is conceived. Here the fundamental or central principles are not understood to be necessary truths or axioms from which other truths are derived. Nor are they considered to be true a priori, in the sense of being known to be true independently of experience. In both of these regards a distinction is made here between what might be termed the level of science, and the level of metascience, with principles constituting the metascientific or transcendental level, which permeates the scientific. This means that while principles need not be necessarily true, nor even true as a matter of fact, they nevertheless obtain a status in science similar to that of necessary truths due to the fact that unless some meta-level is assumed the scientific level lacks coherence.[41] Similarly, the basic principles are not *known* to be true independently of any experience at all, but are *assumed* to be true independently of any *scientific* experience, since any scientific experience presupposes them; and in this respect they may be termed *relatively* a priori.[42] They are arrived at by philosophical reflection

---

[41] In this regard, cf. Agazzi (1988), p. 19: "Science...cannot be pursued without one's using certain criteria of intelligibility which are prior to the specific tasks it involves. In fact, every advancement of some science which has been presented as a 'liberation from metaphysics' has actually been tantamount to discarding a *particular* metaphysical framework and accepting (often unconsciously) a different one. For example, discarding determinism in quantum physics did not mean eliminating all metaphysical views from microphysics, but simply replacing the 'classical' deterministic metaphysics of nature with a new indeterministic one. Therefore it is much more reasonable to be aware of the metaphysics one has, rather than have a metaphysics without knowing it."

[42] Thus our position is similar to Einstein's, where with regard to the notion of categories he says: "The theoretical attitude here advocated is distinct from that of

on the nature of reality as it presents itself to us in the broadest sense. Thus it may be seen that science is not here conceived to be the one true path to knowledge, or the natural outcome of the sophistication of common-sense knowledge. Rather, it is one path among others, based on particular presuppositions and providing a certain kind of knowledge which may or may not be of greater value than knowledge obtainable by other forms of inquiry based on different presuppositions.

As indicated above, principles underlie both the empirical and the theoretical aspects of science, giving direction to the methodologies of both. In the next chapter we shall look at the empirical aspect, and consider in greater detail the role played by principles in shaping it.

———————

Kant only by the fact that we do not conceive of the 'categories' as unalterable (conditioned by the nature of the understanding) but as (in the logical sense) free conventions. They appear to be *a priori* only insofar as thinking without the positing of categories and of concepts in general would be as impossible as is breathing in a vacuum." (1949a, p. 674).

On the notion of the relative a priori and its relation to science, see e.g. Agazzi (1992), p. 41, Lauener (1992), pp. 47–50, and Paty (1992), pp. 98–104.

CHAPTER 3

EMPIRICAL LAWS: THE SUPERVENTION OF EXPERIENCE

Through the years empiricists and realists have been battling over the question of whether or not science should be restricted to investigating what is observable, with neither party paying virtually any attention to the question whether the observable is to be conceived differently from their respective viewpoints. Phenomenalists have argued that phenomena alone should be treated as actually existing, assuming that the empirical laws of science in fact link phenomena in their sense of the term. And realists, for their part, have advocated that science recognise a trans-empirical realm, without considering whether the empirical realm takes a form more in keeping with their view or that of the empiricists or phenomenalists.

This question regarding the nature of the empirical aspect of science could perhaps be answered by directly examining it, to determine the extent to which it is phenomenalist or realist in nature – but that is not the tack that will be taken here. Rather, we shall attempt to show how the empirical aspect of modern science comes to take the form that it does as a result of the methodological implications of the fundamental principles enumerated in the previous chapter. In this way we hope not only to answer the factual question of whether the empirical aspect of science is more in keeping with one or the other of the realist and empiricist views, but also to explain why it is so.

The principles introduced in the previous chapter were of two kinds: fundamental principles and their explicit refinements or qualifications. Where the fundamental principles constitute the core of the scientific enterprise as a whole, the refined principles constitute the cores of its various sub-parts. Thus the reality to which the fundamental principles are to apply includes that to which the refined principles are to apply. With respect to what they are applied to, however, both kinds of principle provide the most general picture possible, depicting it with regard to its spatial, temporal, substantial and causal aspects.

Now this general picture provided by ontological principles can be made more specific, and it is precisely the concern of scientists in their normal capacity to work towards this end. Thus, assuming similar states to succeed similar states according to a rule, the scientist seeks to specify this rule in particular cases. And since such a rule is a rule for the alteration (and not generation or destruction) of substance, the scientist must ensure that the successive states being investigated are changes in one and the same substance. And finally, assuming changes (of state) to result from contiguous causes, the scientist attempts to determine the nature of these causes in particular instances.

Broadly speaking, we may say that the task involving the uniformity principle is primarily that of the empirical scientist; the task involving the contiguity principle is that of the theoretical scientist; and the task concerning the substance principle is taken up by both, each in their own way. In what follows of the present chapter we shall look at the empirical aspect of science and consider how its methodology has been shaped by the principles of uniformity and substance.

1. THE UNIFORMITY PRINCIPLE AND EMPIRICAL LAWS

The principle of the uniformity of nature, which states that change takes place according to a rule such that similar states are followed by similar states, has a prescriptive implication of singular importance for the methodology of empirical science. Given the principle, scientists' efforts are directed to determining which sorts of states are followed by which, that is, to discovering the precise rule by which change takes place in particular cases. This is done by becoming acquainted with states of such a kind as can be seen regularly to be followed by states of some other particular kind. In other words, in accepting the principle of the uniformity of nature, the scientist is urged to attempt to discover natural laws.

The ramifications of this prescription are manifold. For one thing, it directs attention to states of affairs which can in principle, and more particularly in practice, exist at different points in space and time, and to considering the nature of those hitherto unknown changes which accompany them – i.e. it directs empirical investigation to such states of affairs as can be *replicated*. Also, it implies an active rather than passive investigative procedure which involves *controlling the conditions* under which such an investigation is being conducted. And further, it demands a means of establishing *objectivity* in the determination of whether two spatially or temporally distinct states are sufficiently similar to be expected to be accompanied by similar changes.

Replication

The first step in the meeting of these requirements consists in the provision of a *standard*, i.e. something which makes it possible objectively to *compare* different situations with respect to at least one of their common properties in spite of their being separated in space and time. In functioning as a standard, such an entity must be essentially unique; and in order to be used in the making of comparisons spanning different times and places, it must be non-changing.

These two demands on the standard create awkwardness in its employment however, for if it is essentially unique it would seem that it could only be used at one place at a time; and if it is not to change it could prove to be of the wrong dimensions not only for comparing a property's changing with its not changing, but also for comparing its different ways of changing. These difficulties would be overcome however if it were possible to *replicate* the standard itself. Then the replications could be used at different places at the same time, and they could be given suitable configurations for the making of fine and coarse comparisons of the ways in which a property changes.

Such standards exist in empirical science and are used all the time. They determine the *units of measurement* of modern science, and their replications are nothing other than *measuring instruments*.[1] To

---

[1] As expressed by W. Stanley Jevons: "Instruments of measurement are only means of comparison between one magnitude and another, and as a general rule we must assume some one arbitrary magnitude, in terms of which all results of measurement are to be expressed. ... Hence, whether we are measuring time, space, density, mass, weight, energy, or any other physical quantity, we must refer to some concrete stan-

take a frequently cited example, the present standard in terms of which all measurements of length must be made is 1,650,763.73 wavelengths of the orange-red line of krypton 86. This is the standard metre. Earlier this standard consisted of a particular platinum-iridium bar kept hermetically sealed in Paris.

Thus all measurements are comparisons with a common standard or standards assumed to be non-changing. For this reason empirical scientists do not, for example, use parts of their own bodies in measuring, but turn to something that can be taken to be constant, independently of whether it is being used by one investigator or another.

In one sense such standards are arbitrary, in another conventional. While the choice of original standard is relatively arbitrary, once the choice has been made, its continued use is dictated by convention. Note that such a standard cannot take any form whatever, however, but must be such that it can be used in the construction of situations which reveal laws of nature. Thus a particular statue or painting is not taken as, say, a standard of beauty, for what is beautiful to one person need not be so to another, and there is in any case no way of determining whether the results of employing the standard on one occasion are the same as on another.

Mathematics enters empirical science through the physical combining of a standard with one of its replications. Certain standards may be such that their use is not amenable to mathematical treatment other than as a linear ordering, such as in the case of the various kinds of minerals referred to in Mohs' scratch-test of hardness. To constitute a standard for true measurement, the entity in question must be such that it makes possible the *physical addition* of the property for which it is the standard.[2] Such physical addition, like measurement itself, presupposes the principle of the uniformity of nature, and involves the standard of measurement being in some way combined with one of its replications so as to produce *twice as much* of the property for which it is the standard. Thus, for example, in order for something to constitute the standard for the measurement of length, it must be in some way possible to combine one unit

---

dard, some actual object, which if once lost and irrecoverable, all our measures lose their absolute meaning." (1887, p. 305).

[2] For a more detailed treatment of this point, see Campbell (1920), Ch. X.

length, determined by the standard, with another unit length so as to obtain twice as much length; or if the standard is to be one for the measurement of mass, it must be possible to add one unit to another in such a way as to obtain twice as much mass. The existence of such standards means that the properties whose units they determine can be represented on a ratio scale, and that each such property may more particularly be termed a *magnitude*.

Suitable manipulations with the replications of standards of measurement make possible measurement in terms of fractions of units; and the term "magnitude" may legitimately be extended to include properties whose measurement depends on more than just one standard, and which thus may not be susceptible of physical addition.[3] But in all cases the physical standards on which measurements are based are themselves considered constant with respect to time, place and observer, and it is ultimately on them that the ability to replicate situations in empirical science rests.

Control

While measuring instruments based on physical standards are necessary for the discovery of natural empirical laws, in order actually to establish such laws, i.e. to determine the form taken by the principle of the uniformity of nature in particular cases, the instruments must of course be put to use, and in a particular way.

Their first employment is in the creation of a replicable state of affairs in which the natural law can manifest itself, an effort which, practically speaking, involves the construction of an experimental *apparatus*. More generally, it means creating a situation in which determinate changes can be made in a portion of substance the

---

[3] Campbell distinguishes between two kinds of measurable properties, which might be termed fundamental and derived. Fundamental properties are those which can be measured without measuring anything else, and derived properties are those which cannot. Thus e.g. where length is fundamental, density is derived (from length, via volume, and mass). In the present context we should say that the physical standards underlying all measurement are those which provide the units in terms of which *fundamental* measurements are made. For more on this distinction, see Campbell (1920), pp. 275–277; (1938), pp. 126–127; (1942), pp. 763–765; and Ellis (1966), Chh. V and VIII.

(measured) quantity of which is held constant (in accordance with the substance principle), while the effect of these changes on the properties of the substance is observed. This procedure constitutes controlling the conditions of an experiment. In the simplest case, the quantity of one causally relevant magnitude at a time is varied and measured, while all others are eliminated or held constant; and the result of this variation as manifest in the change of the quantity of another magnitude is measured and recorded. Put in other terms, this is to say that once the parameters of an experiment are fixed, the independent variable is modified and the effect of this modification on the dependent variable is determined.

Since the quantity of substance does not change, the mensural values obtained from the pre-change state can be *equated* with those of the post-change state. Though it normally requires a good deal of preparatory work in order to perform an experiment properly for the first time, once this has been done the equation may be taken to be the expression of a natural law (in accordance with the principle of the uniformity of nature); and such an experiment is repeated only to ensure that no error was made in its performance.[4] Empirical laws in the exact sciences are thus expressed as equations representing relations between measurable properties (magnitudes). And it is through the discovery of such laws that empirical scientists believe themselves to learn some of the unchanging rules according to which natural change takes place, that is, to learn the specific form the uniformity principle adopts in particular cases.

Objectivity

The expressions of empirical laws, in that they specify the form of the principle of the uniformity of nature, are not generalisations about what is the case at different times and places, but specifica-

---

[4] Thus, due to the principle of the uniformity of nature, the situation is drastically different from that depicted on the logical empiricist conception, according to which the accumulation of ever more confirming instances is to add greater and greater support to a law. As expressed by Campbell: "So far as induction is a process at all, it is complete after a very limited number of experiments. The finding of laws from a large number of experiments has nothing whatever to do with what is usually regarded as induction." (1920, p. 213).

tions of what must be the case, independently of time or place. Since anyone can check them by reconstructing the situation in which what they depict is most clearly manifest, they may be considered objective. The expression of empirical laws (in the form of equations) thus provides scientific *knowledge*, which is taken to be knowledge of the *facts*[5] of science. In other words, it is only when an expression has been shown to be that of an empirical law that it is admitted as a scientific fact.

One reason that data expressing individual observations or measurements do not attain factual status is that they can be and often are simply mistaken, e.g. as a result of experimental error. What data constitute rather is *evidence* as to what the scientific facts are.

Such facts are usually manifest most clearly in highly artificial or constructed situations, as described above; but even when not, as in the case of the laws of planetary motion, their scientific expressions are directly applicable only to ideal states of affairs. This is to say that scientific laws,[6] while objective, are *idealisational*.

2. THE SUBSTANCE PRINCIPLE AND EMPIRICAL SYSTEMS

Where the principle of the uniformity of nature urges the empirical scientist to discover natural laws, the principle of substance dictates what those laws concern, viz., that which persists through all changes – the substance being investigated by the science in question. Both principles are intimately related to measurement. The uniformity principle must be presupposed in the performance of any

---

[5] Cf. e.g. Enriques (1906), pp. 68, 70: "Facts subject to conditions, of which we have lately spoken, are commonly called 'laws,' especially when they are stated in a simple and general way. [And we conclude] that we cannot recognise a philosophical foundation for the distinction between 'facts' and 'laws.'" The identification of facts and laws is also made by others, including Poincaré, who speaks of law as being "scientific fact itself" (1914, p. 14), and Campbell, who says: "A 'fact' is . . . a portion of experience which is known to be interconnected by a relation of uniformity." (1920, p. 101).

[6] Though in this work the term "law" is normally intended to refer to uniformities existing in nature, for the sake of simplicity it is also sometimes used in referring to the scientific *expression* of such uniformities. The context should make clear which use is intended however.

measurement, and measurement must be employed in determining what, on the empirical level, is to count as the substance of the discipline. Since substance is neither to come into nor go out of existence, unless it comes into a situation from elsewhere or leaves the situation to go elsewhere, its *quantity* in that situation will remain the same, no matter how its other properties are changed. Thus, from the empirical point of view, substance is that the *measure* of which can be held constant through all other changes.[7] While the quantity of substance does not change, other quantities do. What the constancy of substance guarantees is that the sum of all other quantities, as properties of substance, will *equal* each other before and after an experiment. In this way, the expression of empirical laws by equations implies the existence of a substance for which the laws hold.

In physics, that the measure of which is to be the same at the beginning and end of any experiment is energy,[8] and in chemistry it is quantity of matter; and so we have the application of the substance principle in the form of the principles of the conservation of energy and matter respectively on the empirical level.[9] In the case of energy the situation is conceptually more complicated due to the fact that energy not only plays the role of a substance but also that of a cause, as it involves the operation of forces. Other things being equal, a change in the quantity of force exerted in a particular situation will mean a change in the quantity of energy. By excluding extraneous forces, the experimental physicist tries to create a situation in which energy is conserved, i.e. to isolate a *system*, so that the energy change manifest in the variation in the independent variable can *equal* that manifest in the dependent variable. Should these values turn out not to be the same, the principle of the conservation of energy suggests that all relevant causal parameters have *not* been ex-

---

[7] In a similar vein, cf. Weyl (1949), p. 177: "A measure of quantity must be found according to which the transmitted quantum does not change."

[8] Cf. ibid.: "In this sense energy may also be looked upon as substance;" and d'Abro (1939), vol. 1, p. 333: "From this standpoint energy has the properties of a substance."

[9] Strictly speaking, the quantity of substance need not remain constant in an experiment, so long as its change is measured.

cluded or kept constant, and thereby affords a check on whether the experiment has been properly performed.[10]

The substance of a discipline can exist in different forms, which are distinguishable on the basis of the laws that hold of them. Thus, for example, in chemistry the substance is matter, but matter exists in various forms (the various elements), for each of which different laws hold. And on the basis of determining which laws hold for a particular (isolated) form of substance, the empirical scientist can distinguish that form from others.[11]

Thus we might say that the isolation or closing of a system constitutes an essential ingredient in the construction of such ideal states of affairs as are reflected in idealisational empirical laws. In certain cases however relatively isolated systems may exist in nature, thereby relieving scientists of having to create them themselves. This is so, for example, as regards the solar system, where the laws of planetary motion may be obtained by regarding each planet together with the sun as constituting a conservative system. Here the principle of the conservation of energy applies without further ado.

3. CONTINUITY

In that one and the same magnitude as is represented in the expression of one law also appears in that of others, the empirical laws of science as a whole constitute a highly interconnected network, and definitive alterations in the form of any one law will have repercussions affecting the form of other laws. Due to the principle of the uniformity of nature however, such modifications, when they occur, are virtually always ones of refinement; and, while the stock of empirical laws is continually being increased, there is never a drastic change in the network as a whole. Laws that were discovered three

---

[10] It would be exceedingly difficult if not impossible to depict experimentation without bringing in causal notions, thus implying that empirical laws have a causal aspect. What distinguishes laws from theories in this regard is theories' striving to meet the contiguity principle. But for the time being we shall consider such causal notions simply to be 'intuitive,' thereby allowing the empiricist that the expressions of empirical laws, considered by themselves, are essentially formal in character.

[11] For the relevance of this point to the notion of natural kinds, see Chapter 7 below.

hundred years ago, or two thousand years ago, are still considered to hold today, within certain mensural limits – limits which might well have been recognised at the time of the original formulation of the laws. Thus, given suitable refinement, they continue to have a place in modern empirical science.[12] New laws added to the network are generally in keeping with those already present, and are intimately related to them. Thus where revolutions may well occur within modern science, it is not at the level of empirical laws that they do so.[13] Rather, the facts expressed by empirical laws are accumulated over time, giving modern science a continuous aspect; and the network as a whole, in that it consists of specifications of the uniformity principle, may be said to *constitute* that principle to the extent that it has been empirically specified.

4. NECESSITY AND UNIVERSALITY

The notion of an empirical law would seem to involve a certain tension, for it includes the concept of law, which implies some kind of necessity, as well as that of empirical, which appears to pertain to what is merely 'contingent.' Here we see however how the two con-

---

[12] Cf. Campbell (1923), p. 38: "A law is not materially altered in character or simplicity if the values of the magnitudes change, so long as the nature of the relation between them is unaltered. Thus, if a re-examination of the law relating the volume and mass of silver led to the belief that the density of silver is 10.9 and not, as we now believe, 10.5, we should not regard the law as altered materially for the primary purposes of scientific inquiry; but we should regard it as altered, if we found that the mass was not proportional to the volume and therefore that there was no such thing as density."

[13] In this regard, cf. Boltzmann (1904), p. 160: "Every securely ascertained fact remains for ever immutable; at most it can be extended or complemented by the arrival of new items, but it cannot be entirely overthrown. This explains why the development of experimental physics proceeds continuously without any leaps that are too sudden, and why it is never visited by great revolutions or commotions. It is very rare for something to be regarded as a fact and afterwards be found to have been erroneous, and even when it does happen the error will soon be cleared up without this greatly affecting the edifice of science as a whole."

This of course refutes Popper's falsificationist conception as applied to empirical laws. For a general critique of Popper's philosophy of science, see Dilworth (2007).

cepts fit together. Empirical laws *are* necessary in being specifications of the principle of the uniformity of nature, which, like the other principles of modern science, is itself necessary to the enterprise as a whole. But such laws are necessary *only* to this extent, for they could have been otherwise (the uniformity principle could have been differently constituted) without this jeopardising the viability of the principle, or thereby the core of modern science.

Not only does the uniformity principle provide empirical laws or scientific facts with their necessity, but it also provides them with their universality. By denying that location in space and time *per se* are relevant to the unfolding of a process, the rules of change embodied in the facts of modern science are to be considered valid everywhere and always. This is a somewhat misleading form of expression however, for in this context what is intended is that space and time 'collapse.' Thus modern science, even on the empirical level, is not concerned with reality at-a-time[14] or at-a-place, with the aim of generalising this knowledge to all times and places. Rather, it is concerned with knowledge regarding reality independently of time and place. In this way scientific laws may be considered as actually expressing particularities rather than generalities, i.e. facts regarding the particular form taken by the uniformity principle in the empirical world.[15]

5. DISCOVERY, PREDICTION AND TECHNOLOGY

The equation or equations expressing an empirical law depict the rule according to which the magnitudes involved in changes taking place in a particular kind of system are related. In order to determine such laws the requisite system must either exist in nature or be created. When systems are created which have never existed before, the

---

[14] In this regard cf. Northrop (1931), p. 7: "[T]he priority of eternity means that we do not come to nature perceiving it at an instant in an infinite time series; we observe it as something which is eternal first, and come upon the discovery of temporality in its parts later."

[15] For this reason, and others that may be gleaned from what has been said above, it is misleading to conceive of the expressions of natural laws as true generalisations as can be represented in the predicate calculus.

scientist is in the position of possibly *discovering* a law of nature, that is, of determining for the first time the equation or equations describing the interrelations of the magnitudes in the system. The creation of such a system, via the construction of an apparatus, involves *technological advance*, and we thus see that scientific discovery is in fact dependent on such advance.

When the conditions necessary for the manifestation of a particular empirical law are known to obtain, and known quantitative changes take place in certain of the magnitudes related by the equation(s) expressing the law, the scientist is often able to determine the nature of the quantitative change in other magnitudes in the system. And when the quantity of certain of the system's magnitudes (independent variables) can be varied, scientists can themselves bring about the change (in the dependent variables) resulting from such a variation. In other words, the expressions of empirical laws constituting the uniformity principle make it possible to *predict* changes in closed or isolated systems, and in certain cases to *control* those changes.[16]

This latter point is of key importance for technology, for technology and technological development are dependent upon one's ability to control states of affairs in just this way. And so not only does the discovery of empirical laws rely heavily on technological advance, as indicated above, but the converse also holds, whereby exact knowledge of the relations among the magnitudes of substances makes technological advance possible, and the development of empirical science and technology thus proceed hand in hand. But just as the discovery of particular empirical laws may await the creation of a theory suggesting in which states of affairs such laws might be manifest, so may technological innovation depend on the creation of theory suggesting the causal relations underlying the empirical phenomena to be controlled.

---

[16] Cf. Campbell (1920, p. 69) who, in the same vein, says: "It is, of course, invariability in this sense which gives to science its practical value; it is because the connections between observations established by science are invariable that they can be used for prediction; and it is the power to predict that gives the power to control."

6. THE SUPERVENTION OF EXPERIENCE

What kind of knowledge does the knowledge of empirical laws constitute? The expressions of empirical laws tell how a change in the value of one measurable property of a substance will affect the values of other of the substance's measurable properties. Such relations are determined by *measurement*, and so we might first ask: in what does mensural knowledge consist? More particularly, does it differ essentially from phenomenal knowledge, and if so how?

What Is Phenomenal Knowledge?

To determine whether mensural knowledge is distinct from phenomenal knowledge, the nature of the latter must also be made clear, and in such a way that the notion of such knowledge does not presuppose realism. Common-sense realism invokes a form of phenomenal knowledge when it accepts, for example, that seeing something under appropriate circumstances provides knowledge of properties of the thing itself, such as its colour or shape. But for the empiricist, the knowledge acquired under such circumstances cannot be considered knowledge regarding the thing itself. While the empiricist need not deny the *existence* of things in themselves, what he or she must deny is that the *knowledge* we are capable of acquiring is knowledge of things in themselves, i.e. knowledge of their properties as their properties. Thus for the empiricist the notion of phenomenal knowledge must be that of knowledge of phenomena independently of whatever may or may not underlie them in the real world.

Furthermore, phenomenal knowledge is knowledge acquired either through the senses or via introspection. Though the intellect may partake in the gaining of phenomenal knowledge, that of which knowledge is had is an object of sensation.

Quantities vs. Qualities

Regarding mensural knowledge, one of its striking aspects is the fact that it is always expressed in terms of the quantities, i.e. numerical values, of certain magnitudes. Thus one determines by direct or indirect measurement the numerical value of e.g. the length or mass or density of a physical object. Phenomenal knowledge, on the other hand, in being sensory, is not inherently quantitative. Such knowledge is rather, at least paradigmatically, that of the *qualities* of phenomenal objects, such as their colour, or smell, or taste.

Direct perception is normally not quantitative, and it may even be argued that it is never so. Even in the simplest case where one sees two of something, *that* there are two is not part of what is perceived, but requires *abstracting* from the phenomenal situation. This of course is not to say that phenomenal knowledge cannot be quantified (e.g. by counting), but even this is not possible to the extent that the relevant phenomena can be properly measured, i.e. represented on a ratio scale.

Against this view it has been argued, for example, that colour can be measured through determining a particular light's wavelength. But the problem with this line of thinking is that the results of this process do not provide knowledge of the colour: a blind person with the appropriate apparatus might be able to determine the wavelength of a particular light,[17] but that person would not thereby know the relevant object's colour in the phenomenal sense, which knowledge can only be obtained by direct experience. What one measures is in fact the wavelength of the light, which *corresponds* to the colour one would see if one were to look at a surface reflecting the light. But knowledge of the wavelength and knowledge of the colour are two separate things.

Another potential counterexample in this regard involves geometrical notions. One can look at, say, a sheet of paper and see a rectangle; or, one can determine the paper to be rectangular on the basis of measurement. But note that what one measures is not the shape as seen. This is clear when one considers that the phenomenon seen is 'rectangular' only when the paper is looked at straight-on. What is

---

[17] In this regard cf. Poincaré's complementary comment with regard to *force*: "Everything which does not teach us how to measure it is as useless to the mechanician as, for instance, the subjective idea of heat and cold to the student of heat. This subjective idea cannot be translated into numbers, and is therefore useless; a scientist whose skin is an absolutely bad conductor of heat, and who, therefore, has never felt the sensation of heat or cold, would read a thermometer in just the same way as anyone else, and would have enough material to construct the whole of the theory of heat." (1902, p. 106).

measured is something different, for the result of the measurement suggests it to be rectangular no matter how it appears.

A second criterion for distinguishing mensural from phenomenal knowledge is that phenomenal knowledge is always bound to a particular sense. Sights can only be seen, odours only smelt and sounds only heard. Mensural knowledge is not so bound, however, for in many cases it is possible to obtain knowledge of the quantity of a magnitude using different senses. Thus one can measure the length of something using sight or touch, or even sound (in combination with sight or touch) if one were to consider using, say, a roller that clicked at regular intervals.

Another distinguishing feature is that mensural knowledge can be transmitted in a way that phenomenal knowledge cannot. If one understands what is involved in weighing with a balance one will have acquired new knowledge upon being told that an object with which one is not directly acquainted has a mass of ten grams. But, to use John Locke's example, no amount of understanding will help one attain knowledge of the taste of a pineapple upon having that taste described. One must be directly acquainted with the object of phenomenal knowledge in order to have that knowledge, while this is not so in the case of mensural knowledge.

More generally, we may say that where phenomenal knowledge is knowledge *of* something, i.e. is knowledge by acquaintance, mensural knowledge is knowledge *that* something is the case, i.e. is abstract knowledge. When a person knows, e.g., the length of something, they strictly speaking do not know *of* either the thing or its length, but *that* the thing has that particular length. This difference between mensural and phenomenal knowledge in fact explains why, via language, it is possible to transmit the former but not the latter.

A fourth point distinguishing mensural knowledge from phenomenal knowledge is that the former but not the latter necessarily involves the performance of particular *operations*.[18] Of course it is true that even to perceive a colour there must be some motion of the perceiver's eye, but this differs from the operations involved in measur-

---

[18] See Agazzi (1977), pp. 162–164 and 167–168 for a description of how the operations performed in an empirical discipline are linked to objectivity and replicability, and serve to construct the objects of the discipline.

ing in that it does not necessarily involve the use of *instruments*, as do all measurements other than mere counting.

Against this background we can thus conclude that mensural and phenomenal knowledge are fundamentally different, and that as a consequence the knowledge afforded by empirical laws is not knowledge of the relations between phenomena in a phenomenalistic sense. But we may still ask why this is so, and further pursue our question as to the nature of the knowledge of empirical laws.

On the Distinction between Primary and Secondary Qualities

As a next step in this endeavour we might give some consideration to the distinction between what have been termed primary and secondary qualities. This distinction has been made in various ways since it was first enunciated by the early atomists,[19] but perhaps the way most closely related to present considerations is that of Democritus himself, according to which primary qualities are properties of things in themselves, and secondary qualities are the sensations we experience when we come in contact with things in themselves. Thus secondary qualities depend on a perceiving subject for their existence while primary qualities do not.

The distinction between primary and secondary qualities was used already by Democritus in his explanation of disagreements arising out of sense perception. That honey should taste sweet to one person and bitter to another depends on their differing bodily states. Since honey in itself cannot be both sweet *and* bitter, Democritus concluded that it was neither – that sweetness and bitterness were qualities resident in the perceiving subject, not in things in themselves. He thus considered knowledge of secondary qualities, obtained via the senses, to constitute 'bastard' cognition, while knowledge of the primary qualities of things constituted 'legitimate' cognition.

According to Democritus the way one was to attain legitimate cognition was via a priori reasoning concerning the properties of the

---

[19] It is noteworthy in this regard that some form of the distinction was accepted by virtually all of the scientifically most influential thinkers of the seventeenth and eighteenth centuries, including Bacon, Kepler, Galileo, Descartes, Hobbes, Boyle, Newton and Locke. For a discussion of the historical roots of the distinction and its influence on the founders of modern science, see Dilworth (1988).

smallest constituents of matter, or atoms. This led to the conception of primary qualities as being those properties all bodies must have by virtue of their being bodies. The list of such properties was thus rather short, normally taken to consist of size and shape, and sometimes weight. The secondary qualities, on the other hand, included sights, sounds, smells, tastes and sensory feelings.

The listing of the primary qualities does not tell us however which properties are operative in which situations, and in what way. In an attempt to provide such information, Democritus claimed that, for example, our experience of sweetness is due to what is experienced as being sweet consisting of round and not particularly small atoms (coming in contact with a suitably formed tongue and palate), while a sharp taste is produced by small atoms with many corners. It is not clear whether Democritus considered this form of reasoning to be part of legitimate cognition; but in any case we note that, apart from its being mistaken, its hypothetical nature keeps it from deserving to be called knowledge.

Can no knowledge then be had of the particular primary qualities of anything, that is, of the properties of things in themselves? Kant clearly considered the obtaining of such knowledge to be impossible. But what of the empirical laws of science? They do not constitute knowledge of sensory phenomena or the relations among such phenomena. Do they perhaps constitute knowledge of primary qualities in Democritus' sense – in spite of Kant's views on the matter?

Objective Standards and Substance

One thing that speaks in favour of so viewing empirical laws and the measurements on which they are based is that they meet Democritus' criterion of receiving universal assent. There are no disagreements regarding the results of measurement that are not soon settled, making the knowledge they afford intersubjective.

A second point is that once accepted the expressions of empirical laws retain their factual status through the development of science. They do not concern transitory states – as are phenomenal states – but a realm which in a certain regard is unchanging through time.

Now these two points also apply to formal knowledge in logic and mathematics, which we would not want to claim to constitute knowledge of the kind of reality of interest here. One difference between formal knowledge and the knowledge provided by measurement however is that the latter is ultimately based on standards conceived of as existing in the real world. Empirical laws are determined by comparing states of affairs with such standards, states of affairs which must also be conceived as existing in the real world in order for such a comparison to be meaningful. Furthermore, similarly to the constancy of the standards on which knowledge acquisition is based, that about which knowledge is acquired – the substance the law concerns – is also conceived as independently existent and as constant through time.

As a consequence of these considerations then, we may claim that not only is the knowledge of empirical laws not knowledge regarding either phenomena or a purely formal realm, but, given the relevant principles, it has every right to be considered knowledge of things in themselves,[20] i.e. knowledge regarding the substances for which the laws hold.[21] It is to be borne in mind however that this

---

[20] In this regard cf. Blackmore (1982), p. 79: "Galileo regarded the *purpose* of 'the attempted reduction of scientific experience to experience that can be expressed in mathematical terms' to be to understand the real physical world *beyond* sensations, consciousness, and 'experience' and that what science is 'about' is not just 'experiential possibilities expressed in mathematical terms,' but trans-experiential *physical realities.*"

[21] Cf. Northrop (1931), p. 132: "If nature is nothing but a group of sense data then experimental procedure is pointless." Contrast with Duhem: "[E]xperimental method...is acquainted only with sensible appearances and can discover nothing beyond them." (1906, p. 10).

Regarding Kant, see Prichard: "It is all very well to *say* that the substratum is to be found in matter, i.e. in bodies in space, but the assertion is incompatible with the phenomenal character of the world. ... Now Kant, by his doctrine of the unknowability of the thing in itself, has really deprived himself of an object of apprehension or, in his language, of an object of representations. For it is the thing in itself which is, properly speaking, the object of the representations of which he is thinking, i.e. representations of a reality in nature; and yet the thing in itself, being on his view inapprehensible, can never be for him an object in the proper sense, i.e. a reality apprehended. ... Kant is in fact only driven to treat rules of nature as relating to representations, because there is nothing else to which he can regard them as relating." (1909, pp. 273–274, 280–282). See also Caird: "Kant does not in any way

does not imply that such knowledge can be had of all types of things in themselves, nor that it reveals everything regarding those things about which it can be had. As regards the first point, it would appear that such knowledge cannot be had of the spiritual aspect of reality, whether it be knowledge of the self, or phenomena experienced by the self, or such entities as spiritual forces, should they exist. As regards the second point, the knowledge of empirical laws acquired in the specification of the uniformity principle consists only of quantitative comparisons to a standard, which implies that such knowledge concerns only the *form* of its subject, i.e. concerns the structure but not the content of reality. Furthermore, it is to be noted that the principles such knowledge presupposes, namely those of uniformity and substance, ought not be taken to constitute an apodeictic a priori, but only a relative one.

Two Senses of the Term "Empirical"

The above considerations lead one to realise that the word "empirical" has at least two distinct senses. The one sense, which we might term the phenomenal, is well captured in the *Concise Oxford Dictionary*, where "empirical" is defined as 'regarding sense-data as valid information.' The dictionary provides another definition however, which is much closer to the sense implied when speaking of empirical laws, viz.: 'based or acting on observation or experiment, not on theory.' While this sense may be acceptable in itself, to obtain the particular meaning of "empirical" that has occupied us in the present chapter it might nevertheless be refined, and be taken more particularly to be that of *mensural*, or, *having to do with measurement*.

The distinction between the phenomenal and mensural senses of "empirical" has been little recognised by philosophers, but it is one which is vital to an understanding of the empirical aspect of modern science. It is *measurement* that lies at the heart of modern empirical

---

attempt to show how the idea of matter is derived from experience, except by saying that it is by motion alone that outer sense can be affected." (1877, p. 492).

science, not the experience of sense-data, nor even observation in the ordinary sense of inspecting with the naked eye.[22]

This is not to deny however that measurement and the determination of empirical laws involve phenomenal experience,[23] but to emphasise that the realm investigated in this way is not the ever-changing world of phenomena but a non-changing aspect of the real world. The knowledge scientists gain of empirical laws through measurement and the experimental method, presupposing the principles of uniformity and substance, is not knowledge of their own experiences of the world, but of the world itself. In present terminology this is to say that while empirical science would be impossible without phenomenal experience, the objective results it obtains in its specification of the uniformity principle are the results of the *supervention* of just that experience.

Other Instruments

We have been led to concentrate on the knowledge of empirical *laws* in our examination of the empirical aspect of modern science as a result of the methodological implications of the principle of the uniformity of nature. In this way we hope to have both established that and explained why laws are central to the empirical aspect of science. But this aspect of science involves other kinds of knowledge as well, most notably, knowledge of the existence of particular entities or kinds of entities.

The acquiring of such knowledge often involves the use of instruments, such as telescopes and microscopes, whose primary function is not to measure but to extend the senses. Our claim in this regard then is that such activities might well constitute a part of modern science, but they are more peripheral than is the establishment of

---

[22] This serves to distinguish modern science from Aristotelian science, for, as noted by Louk Fleischhacker, where in modern science one knows things by their measure, in Aristotelian science one knows them by their appearance.

[23] "Thus, in order to understand what physical properties such as mass or elasticity are, it is certainly necessary to have had some experiences, simply in order to be conscious and capable of knowing and understanding anything at all. There is, however, no particular sort of experience that it is necessary to have had." N. Maxwell (1984), p. 202.

laws. The use of telescopes and microscopes, unlike measurement, is not peculiar to modern empirical science, and might with just as much right, or perhaps even more, be employed e.g. in Aristotelian science. But in any case, even the knowledge afforded by such instruments is not phenomenal knowledge, but presupposes realism. A hitherto unknown planet discovered with the aid of a telescope, or a new kind of bacteria seen through a microscope are conceived to exist independently of the investigator and to occupy the real world. However, such knowledge supervenes experience only if its acquisition involves comparison to an independent standard, a comparison which in principle can be performed using different senses. The construction of telescopes and microscopes of course involves such comparison, but their use need not. Thus in their case it would perhaps be best simply to say that the knowledge acquired involves an *extension* of experience, but experience of the world, not of mere phenomena. Moreover, the acquisition of such knowledge is invariably followed directly by attempts to obtain mensural information regarding the entities in question, which can allow the determination of laws regarding them.

Principle-Ladenness

That modern empirical science is not based on phenomenal experience is becoming generally recognised in today's philosophy of science, and is finding expression in the claim that all scientific terms are *theory-laden*. Those who speak in this way seldom distinguish between theories and laws however, let alone between theories, laws and principles, and so one cannot be sure that what they hope to express is not something different from what they actually say.[24] Furthermore, the fact that it is *terms* that are thought to be theory-laden, and not, say, concepts, suggests that in spite of their rejection of phenomenalism these philosophers adhere to what is fundamentally a logical empiricist conception of science.

What we should suggest on the other hand, as is clear from the considerations of the present chapter, is not that all terms of empiri-

---

[24] For a criticism of Karl Popper, N. R. Hanson and Paul Feyerabend in this regard, see Dilworth (2007), pp. 138–140.

cal science are theory-laden, but that all scientific *concepts* are *'principle-laden'* – that is, that the whole of modern science presupposes certain principles, and as regards its empirical aspect more particularly, that it presupposes those of uniformity and substance.

7. EMPIRICAL LAWS REQUIRE EXPLANATION

The methodological prescriptions of the principles of the uniformity of nature and of substance suggest the performance of operations leading to the discovery of empirical laws through the suitable use of measuring instruments and apparatus constructed with the help of such instruments. In conjunction these two principles guarantee that the results of similar operations on an isolated substance will themselves be similar. But neither principle dictates what these results will be, nor suggests how they are brought about.

Once the results have been obtained, that is, once a particular law has been established, the question thus arises as to how it comes to take the form it does and not some other. Given the basic principles enumerated in the previous chapter, the answering of this question consists in indicating how the form assumed by the law is the result of the uniform operation of a certain *cause* or causes, as well as in showing how the causes operate in such a way that they are *contiguous* with their effects. To achieve this end is to explain the law; and the provision of such an explanation is the task of scientific theory.

CHAPTER 4

SCIENTIFIC THEORIES: CLOSING THE CIRCLE

Though measurement and the discovery of empirical laws presuppose the adoption of an epistemological realism, they leave untouched that aspect of reality which empiricists have been most concerned to avoid, namely causes that are actually productive of their effects. In the belief that such causes are truly unknowable, empiricists have either forbidden their investigation, or replaced them with 'causes' that are simply manifestations of the principle of the uniformity of nature, i.e. constant conjunctions. The aspect of modern science we shall consider now is precisely that which investigates the nature of real causes, and it is thus fundamentally realist in its orientation. In its paradigmatic form it involves the construction of theories intended to explain empirical laws by indicating both the regular causes underlying them, as well as how those causes operate in such a way as to be contiguous with their effects.

In the previous chapter we saw how the principle of the uniformity of nature and the principle of substance set guidelines for the doing of empirical science. The uniformity principle directs attention to the creation of replicable states of affairs, while the substance principle to a large extent delineates the way in which such states of affairs are to be investigated. This methodology leads naturally to the determination of empirical laws, which constitute the specific form taken by the uniformity principle in empirical situations. It does not, however, lead to the explanation of such laws, that is, to the indication of the nature of the causal relations that underlie them.

Where the expressions of empirical laws are specifications of the principle of the uniformity of nature, we shall find in the present chapter that scientific theories are attempts to explain empirical laws by specifying the principles of substance and causality in such a way that the contiguous nature of the causal relations underlying the laws is indicated.

1. THEORETICAL REDUCTION AND THE CLOSING OF THE CIRCLE

On the conception of science being presented here, principles provide a general picture of the reality a science or scientific discipline investigates, while theories and the expressions of laws consist of applications of the principles to that reality in such a way that its nature is indicated in greater detail. There is an asymmetry however in how this is done in the case of laws as compared to theories. As is indicated by the structure of the present work, the establishment of an empirical law or laws is to be understood as being conceptually prior to the propounding of a theory. The purpose of theories being to explain laws, the laws must be assumed to be known. This conceptual priority however does not imply a temporal priority, it being quite possible that a theory be formulated prior to the discovery of the law or laws it explains.[1]

Following the prescriptive implications of the uniformity principle, the empirical scientist creates replicable states of affairs in which natural laws are manifest. That such laws exist is assumed in adopting the uniformity principle, but nothing is assumed regarding the particular form they take, other than that it will be in keeping with the principles as a whole. How it in fact is so, however, is something requiring demonstration. Thus we may picture the situation as one where the scientist begins on the level of principles – the uniformity and substance principles – then reaches out into empirical reality and establishes some facts which presuppose those principles, and then is obliged to show that these newly won facts can themselves be assimilated on the level of the principles – more particularly, by the principles of substance and causality. The task of indicating how this can be done is that of theories, and in this way we

---

[1] In this regard, see Whewell (1847), Part 1, pp. 651–652: "*The Study of Phenomena leads to Theory.* – As we have just said, we cannot, in any subject, speculate successfully concerning the causes of the present state of things, till we have obtained a tolerably complete and systematic view of the phenomena. Yet in reality men have not in any instance waited for this completeness and system in their knowledge of facts before they have begun to form theories. Nor was it natural, considering the speculative propensities of the human mind, and how incessantly it is endeavouring to apply the Idea of Cause, that it should thus restrain itself."

can see theories as closing the circle begun by the establishment of laws by drawing it back to the level of principles.[2]

Much has been said about *reduction* in speaking about science, the dominant logical empiricist conception being that according to which theoretical notions are to be reducible to empirical ones via 'reduction sentences.'[3] Here we see however that reduction actually operates in quite the opposite direction. As far as modern science is concerned, in the case of *concepts* reduction consists in the expression of all empirical (mensural) concepts in terms of notions found in the underlying principles, particularly in their refined form. In other words, notions which originally are only mensural in character – such as 'temperature' – are shown to be completely expressible in terms of the categories delineated by the principles underlying the science in question – such categories as 'mass' and 'motion.' In the case of *laws* or *facts*, scientific reduction may be said to consist in their being demonstrated to be but manifestations of the principles on the empirical level. Both kinds of reduction are accomplished by the propounding of scientific theory.

If we look more closely at how this is brought about we see that while the key principle to which reduction is made is a refined form of the (contiguity) principle of causality, a refined form of the principle of substance serves as the mediator in the reduction.

2. THE SUBSTANCE PRINCIPLE AND THEORETICAL ONTOLOGIES

The equations expressing empirical laws afford algorithms for determining the numerical values of certain empirical properties given the values of certain other such properties, but they do not by themselves indicate the nature of the causal relations underlying the laws

---

[2] Thus our conception of theory is very much in keeping with the notion expressed in the *Concise Oxford Dictionary*, viz.: "Supposition or system of ideas explaining something, esp. one based on general principles independent of the facts, phenomena, etc., to be explained."

[3] The notion is Carnap's. For a concise presentation of this approach, see Hempel (1965), where it is admitted that "reduction sentences do not seem to offer an adequate means for the introduction of the central terms of advanced scientific theories" (p. 110).

they express. Theories constitute attempts to specify such relations by depicting realities in which they are conceived to be operative. When a theory succeeds in indicating the contiguous nature of the causal relations underlying a particular empirical law, it has explained that law.

Other things being equal, the reality in which causal relations are operative could consist solely of causes, its substantial aspect being the constancy of these causes or 'causal powers'[4] through time. In practice however, for reasons to be taken up in the next chapter, theoreticians have preferred to include the independent conception of substance as the bearer of causes.

The reality depicted in a scientific theory may be intended to be precisely the same reality as that in which the empirical law or laws of interest are manifest; otherwise it may be only a part of that reality, if only a part is conceived of as being causally relevant to the law or laws whose causal underpinning is to be indicated. In either case, just as the substance being investigated in the empirical situation must be isolated, so must it in the theoretical; and the principle of substance is to apply to both situations. The main difference however is that where in the empirical situation the substance is distinguished only by its mensural constancy, its components consequently being undifferentiated, in the theoretical situation it is further distinguished by its ability to bear causes, which leads to its components virtually always being differentiated. The reason for this is that such a differentiation is required in order that the causes borne by the substance can be contiguous with their scattered effects while at the same time the substance neither comes into nor goes out of existence. The conception of a substance with differentiated parts is thus a representation of those entities the existence of which is necessary for the discipline. This representation may be termed an *ontology*, an ontology consisting of notions representing *ontological entities*.

---

[4] The idea of causal powers can be traced at least as far back as to Leibniz, the substantiality of whose monads is identified with their power. Boscovich's theory of matter was close to being a pure causal-power theory, the only 'substantial' entities being points of force. For a relatively recent philosophical treatment of causal powers, see Harré & Madden (1975).

The most straightforward way in which ontological entities may be represented is as themselves neither coming into nor going out of existence – at least for the duration of the phenomena the theory is intended to explain. This conception is the most common and is of especial relevance to the explanation of natural kinds, as will be taken up in Chapter 7 below. That ontological entities be relatively permanent is not necessary however, so long as what becomes of an extinguished ontological entity is immediately another such entity belonging to the same substance. Thus, for example, if the substance is energy, and mass is taken to be a form of energy, then particles (ontological entities) having mass can cease to exist so long as they are directly transformed into another sort of energy of the same quantity. Also, though it is perhaps most natural to conceive of ontological entities as separate individuals, this is not necessary either, and they may instead be taken to be components of a continuum, in the event that the substance is conceived e.g. to constitute a field.

But no matter how the substance is conceived, it must conform to the refined principle of substance of the discipline. Thus, in the case of physics for example, it must be a form of energy; or in the case of chemistry, it must be a form of matter. As the primary function of the substance is to be the bearer of causes, the ways in which ontological entities relate must likewise be in keeping with the discipline's principle(s) of causality.

3. THE CAUSALITY PRINCIPLE AND CAUSAL MECHANISMS

The notion of causality employed in modern science most probably has its origin in the experiences we humans have of moving things or of being moved by them.[5] Thus the effort we expend in lifting a

---

[5] This can be distinguished from the origin of the (Aristotelian) teleological notion of causality, which, it may be argued, lies in the propensity of living organisms to *grow*. It should also be distinguished from the origin of the (Humean) positivistic notion, which lies in the passive observation of change of motion. For others' considerations on the origins of the modern-scientific notion, see Fröhlich (1959), where he suggests that "if we list those factors which are necessarily involved when we move objects, we should have a roughly accurate selection of those properties in terms of which we can understand the operations of bodies on each other or by

stone or throwing an opponent in a wrestling match, or the pressure
we feel when buffeted by a stiff breeze or when jostled in a crowd
are experienced by us to be causes having effects either on ourselves
or on other things. We then extrapolate this notion of causality to
situations where we do not experience effort or pressure, and speak
of physical objects' having causal relations among themselves, such
that an earthquake can cause a building to collapse, or one billiard
ball's colliding with another can cause that other to move.

In each of these cases, rather than the general notion of causality,
we could apply the more particular notion of force; and it may be
suggested that for modern science the paradigmatic notion of causality is in fact that of force as the cause of motion (acceleration).[6] But
more generally, we may say that modern theoretical science consists
essentially in the indication of how particular empirical regularities
result from the regular operation of certain causes between ontological entities. Thus one way of characterising a scientific theory is
to say that it constitutes the attempt to present that causal ontology
whose epistemology consists in specific known empirical laws. This
does not exclude, however, that theoretical scientists in their research
may be more interested in the nature of the causes themselves or the
entities that bear them than they are in how particular phenomena are
produced.

The general nature of the causal relations conceived of in each
theory must be those set down by the particular form(s) of the prin-

---

reason of which we say objects act on each other. In all such situations we exert force
against a body which, if stationary, resists our efforts to move it, or, if in motion,
resists our efforts to stop it." (pp. 212–213); and Weyl: "In our will we experience a
determining power emanating from us, and were we not thus actively and passively
drawn into the stream of nature (be it even merely in the role of an experimenter who
creates the conditions of the experiment), we would hardly regard nature under the
metaphysical aspect of cause and effect." (1949, p. 192).

[6] In this regard cf. Campbell (1920), p. 156: "[A] change of colour, of pitch, of electric
charge is not a change of motion, so far as experiment can determine. But if [change]
is produced voluntarily it is always a result of motion produced by force; .... And
though we cannot express this relation in our laws, we have an opportunity to do so
in our theories. By framing a theory in which the hypothetical ideas are concerned
with motion and force, and in which the dictionary relates these ideas to colour, pitch
or charge, we can establish that these changes are 'really' the effects of change of
motion."

ciple of causality adopted in the science. These forms of the causality principle themselves invariably involve the principle of uniformity, such that similar causes produce similar effects. So, for example, in the kinetic theory representing an ideal gas, the causal relations among the molecules constituting the substance accord with Newton's three laws (causal principles) of motion, which themselves express a uniformity in causal behaviour.

That a theory depict uniformly operating causes however is not in itself sufficient for it to provide a completely satisfactory explanation of the phenomena in which the causes are considered to be operative. What more is required is an indication of *how* the causes bring about their effects; and as conceived in modern science this question cannot receive a wholly adequate answer so long as the causes and their effects are thought to be separated either in space or in time. In other words, a satisfactory theoretical explanation of phenomena must indicate how they arise as a result of the operation of causes which are *contiguous* with their effects.

In the case where the theory in question succeeds in depicting a regular cause operating via the substance in a way which is in keeping with the contiguity principle, that part of the substance which mediates the causal relation may be termed a *causal mechanism*. And we may say quite generally that it is the contiguity principle that lies behind the search for causal mechanisms in science, as well as behind scientific theorising as to their nature.[7] And it is for lack of the fulfilment of the contiguity principle that, for example, scientists still today consider themselves not to have discovered or adequately

---

[7] Cf. Harré (1970b), p. 74: "By and large the sciences do not recognize, and never have recognized, action at a distance. It has seldom been accepted as more than a temporary and unwelcome expedient, at the best a metaphor for ignorance of causes. It is often just when a gap of space and time intervenes between two parts of a process that we invent hypothetical entities or processes to satisfy the principle of the proximity of some cause to an effect." In this regard see also Enriques (1934), p. 25: "[T]he whole development of physics after Newton and up to the most recent conception of relativity tends to frame phenomena, or the reciprocal actions of bodies, as propagating by contiguity in space and time." This has often meant the theoretical conception of either fields or particles as constituting causal mechanisms affording the contiguity of particular causes and their effects.

conceived of the mechanism of gravity, if there be such a mechanism.[8]

4. THE HYPOTHETICAL ASPECT OF THEORIES

As is generally recognised, the adequacy of a scientific theory will at least in part depend on how closely its empirical implications are in keeping with known regularities. Establishing whether this is so is normally a relatively straightforward affair involving the determination of how the theory's ontology should be empirically manifest, and comparing this with the relevant experimental laws. (This procedure will be considered in more detail in the next chapter.) Also of relevance to the theory's acceptability is of course whether the ontology of causal mechanisms in the theory has a counterpart in reality. To determine this is often more difficult however.

Theories, by their very nature, are normally put forward in situations where the cause or causes underlying a particular regular change are not evident. Such a situation might be, for example, one which apparently involves no motion. So the theory must refer to an aspect of the situation which is not directly manifest – which at the time is not accessible to measurement or sense-extending instruments – in order to provide its explanation. Thus theories, unlike the expressions of empirical laws, are virtually always hypothetical when first advanced, and transcend the kind of information available

---

[8] In this regard, see Newton (1687), p. 634: "That gravity should be innate, inherent, and essential to matter, so that one body may act upon another at a distance through a vacuum, without the mediation of any thing else, by and through which their action and force may be conveyed from one to another, is to me so great an absurdity, that I believe no man, who has in philosophical matters a competent faculty of thinking, can ever fall into it." See also Helmholtz (1894, pp. xxxiv–xxxv): "Gravitation still remains an unsolved puzzle; as yet a satisfactory explanation of it has not been forthcoming, and we are still compelled to treat it as pure action-at-a-distance." Consider also Fröhlich (1959, p. 216): "Many scientists and philosophers rejected gravitation as a fundamental property of matter and even rejected the whole Newtonian system of explanation, on the ground that it was inconceivable that one body should act on another at a distance. This rejection was not due to their never having seen material objects attract each other at a distance, for they were familiar with magnets; it was due more probably to their keeping, as an implicit model of how bodies can act on each other, our ways of voluntarily acting on objects."

at the time that they are formulated. They state that *if* the world has such-and-such a particular nature, which it is beyond our present capabilities to determine, *then* the relevant laws follow from the principles as a matter of course. In this way theories reach out into the unknown to explain the known; they attempt to clarify what is revealed in terms of what is not.

Thus, unlike the expressions of laws, theories can often only be tested indirectly.[9] It is to be noted that this aspect of theories is not due to there being a potentially infinite number of empirical cases to which they should apply – as e.g. Popper would have it – since for scientists the principle of the uniformity of nature ensures that if a theory applies to one case it will apply to all similar cases. The situation rather is one where the reality depicted by the theory's ontology is not epistemologically accessible.[10] This inaccessibility of reality means that *competing* theories may be put forward, each with the intention that the workings of its ontology best depict the causal underpinnings of the law or laws in question.

In constructing such ontologies the theoretician is free to *speculate*; so one researcher might depict the discipline's substance as being discrete, while another depicts it as continuous (cf. corpuscular vs. field theories). Or two researchers might agree in depicting the substance as discrete, while differing on how many sorts of ontological entity it contains, or on what the natures of the sorts are. In such cases each theory presents an account of the same phenomena based on exclusive conceptions of that aspect of reality constituting the substance of the discipline. And just as all causal relations depicted in theories of the same discipline must be in accordance with that discipline's principle of causality, so must all theoretical ontologies be in accordance with the relevant principle of substance.

Not every theory need be hypothetical in the above sense however; nor, if hypothetical, need all theories be so to the same degree.

---

[9] Cf. Campbell (1920), p. 130: "But a theory is not a law; it cannot be proved, as a law can, by direct experiment."

[10] Cf. J. C. Graves: "There seems to be no way of verifying or falsifying an ontological hypothesis by seeing whether or not it corresponds with observable facts, since its very nature is to transcend our experiences by exhibiting them as merely the sensible manifestations of an underlying reality." (1971, p. 54).

Furthermore, in the case where the properties of the substance depicted by a theory are inaccessible when the theory is first proposed, they need not remain so. Technological advance can lead, and often has led, to the establishment of the correctness or mistakenness of theoretical ontologies.[11] In such cases where one theory by this or another means comes to replace a competitor as the best candidate for representing the fine structure of the substance of the discipline, we may speak of a *minor* scientific revolution. Such a revolution is minor rather than major since it does not involve an alteration in the principles underlying the science or discipline in question.

Common to all theoretical speculation are the constraints set by the discipline's principles. These constraints may not always be easy to work within however, and the mark of a good theory (at least when it is initially being considered) in the case where the reality the theory depicts is inaccessible is not only how well its empirical implications are in keeping with known empirical regularities, but also how well its ontology is in keeping with the principles of the discipline.[12]

5. EXPLANATION, UNDERSTANDING AND THE LIMITS OF INTELLIGIBILITY

In the event that a theory's causal ontology has a counterpart in reality, and at the same time the expressions of empirical laws can be derived from the theory, the theory may be said to *explain* those laws. It shows how the laws are but a manifestation of the functioning of the discipline's principles in nature. And, rather generally, it may be said that the primary purpose of theories in science is to provide just such explanations. When the contiguity or other causal

---

[11] On the importance of sense-extending instruments in this regard, see Harré (1970b), p. 83.

[12] Thus we do not agree with Galileo if, in his argument with Bellarmine, he is to be taken as suggesting that the *only* criteria for the acceptability of claims about the trans-empirical are empirical. There are other criteria as well, based on the principles of the discipline in question.

principles of the discipline are emphasised, the explanation provided by the theory may be termed a *causal* one.[13]

In such cases as where the theory provides an explanation, those receiving that explanation may be said to obtain *understanding*. They come to understand *why* the laws take the particular form that they do, against the background of the principles of the discipline. Where the contiguity principle is emphasised, they may be said to come to understand *how* the facts come about as a result of the operation of causes which are contiguous with their effects.

However, theories most often are, and might well remain, hypothetical. This means that one cannot be certain that such theories actually do provide explanations, or that what one obtains in receiving such 'explanations' is in fact understanding. In such cases then we might better speak of the theory as providing an *account* of the phenomena, thereby making them *intelligible* to those receiving the account, or affording them with *an* understanding (or a *hypothetical* understanding)[14] of the phenomena. Or we might say, using Campbell's phrase, that such accounts provide *intellectual satisfaction.*

But whether or not a theory be hypothetical, the relation of theories to facts is here clearly distinguished from that of the expressions of laws to facts, the latter not constituting understanding, but *knowledge.*[15] Where we speak of laws or facts as being *discovered*, we do

---

[13] In this regard, cf. McMullin (1984b), p. 210: "Theory explains by suggesting what might bring about the explananda. It postulates entities, properties, processes, relations, themselves unobserved, that are held to be causally responsible for the empirical regularities to be explained." This of course is not to deny that theories may also be of value in other respects, such as in suggesting the existence of hitherto undiscovered laws, or in indicating the common causal basis of previously unconnected laws.

[14] In this regard cf. Boltzmann: "In order to *understand* the phenomena which actually occur, we may draw conclusions from hypothetical assumptions, that is, from processes which, though possible on analogy with similar phenomena in other circumstances, cannot be observed and may not even be observable in the future, owing to their speed or small size or something similar." Cited in Flamm (1983), p. 261, from a lecture given by Boltzmann in 1903; emphasis added.

[15] Basically the same distinction as that being drawn here has also been made by Peter Alexander; in his words: "The aim of explanation is to achieve understanding, to make things intelligible, whereas the aim of description is to say how things are." (1963, p. 138). This distinction is also taken up by d'Espagnat (1992), Fleischhacker

not use the same form of speech in talking about theories. Theories are abstract conceptual schemes *created* or *constructed* to account for the facts. Furthermore, where one sees that knowledge of empirical laws greatly assists technological advance, the understanding provided by theories does so only to the extent that it suggests what the laws might be. Thus just as an understanding of the causes underlying phenomena can be of relevance to the empirical scientist, so can it to the technologist. But in both cases what is of central concern lies on the empirical level. Where the empirical scientist aims to establish the existence of a regularity, the technologist aims to construct an apparatus which functions as it should, and in neither case is the acquisition of an understanding what is being aimed at. And where laws constitute a continuous aspect of science – the body of known laws ever growing and experiencing no serious disruptions – theories, due to their originally hypothetical nature, constitute a discontinuous aspect, where one theory can be replaced by another which has quite a different ontology.

Thus scientific theories, unlike the expressions of laws, provide accounts of various aspects of the empirical world, making them intelligible in terms of the principles of the discipline in which the theory is developed. But it may be the case, and has been the case as a matter of historical fact, that as regards particular phenomena scientists are or have been unable to construct a theory which is both in keeping with the discipline's principles and at the same time capable of accounting for the phenomena. As an example we might consider Bohr's theory of the atom. From the theory one can derive the empirically confirmed Balmer formula representing the wavelengths of the spectral lines of hydrogen. At least when the theory was first introduced however, it faced problems of principle, the one to be considered here having to do with its mechanical depiction of electrons as being able to move from one orbit to another without traversing the intervening space. This requirement contravenes the principle of the continuity of space implied by the uniformity principle; and given the continuity principle, the account the theory provides of spectral lines is notably weakened: how can one claim to understand

---

(1992, p. 249), Ellis (1992, pp. 266–277, 279) and Manicas (1992, pp. 283, 296–297).

a particular phenomenon when its explanation requires one to accept that a physical object can disappear from one place and instantly reappear at another? Lacking an explanation more in keeping with the principles however, the best explanation available is often retained; and what might well happen after some time is that the state of affairs as depicted in the theory lead to a reformulation of the principles themselves. In the present example we can imagine this meaning the replacement of the principle of the continuity of space with a principle stating that space is discontinuous.

Thus we see one way in which the principles of a discipline can undergo change. Another way is through a direct investigation of currently adhered to principles and potential replacements against the background of accumulated scientific fact, as in the case of Einstein's investigations regarding the principles of relativity and constancy of the speed of light.

Recognising the possibility of such changes, it becomes a matter of philosophical interest as to how radical a change of principle(s) would have to be to warrant our saying of a science or discipline that it has become essentially different from what it was before – that it is no longer the same discipline. A generally recognised example of this kind of change is that from Newtonian to quantum mechanics. When such a change occurs we can speak of a *major* scientific revolution.

Given the possibility of such major revolutions the question arises as to how flexible human understanding is as regards its ability to accept the picture of reality depicted by the new set of principles. Principles are the touchstone for understanding *within* a science, but what if they cannot be conceived as coherent in a broader context? The question is not one of which principles are correct or come the closest to being correct, for even 'correct' principles, i.e. principles which lead to the acquisition of knowledge, might depict an incomprehensible reality. This state of affairs might of course be ameliorated with time, as people become more accustomed to the new principles – as has been suggested e.g. with regard to the more counter-intuitive aspects of the theory of relativity. But if such an intellectual transition proves impossible, and the reality the new principles depict remains incomprehensible, then the science in question, in that it cannot itself provide intelligibility in a broader context, will be lacking what may be the most important feature a science can have.

That there is ultimately more to explanation and understanding than that phenomena be shown to be a manifestation of the principles of the discipline in which they are studied – that such principles ought in some way be connected to what people believe themselves capable of understanding in everyday life – is a topic that will be returned to in the next chapter. There, more generally, we shall attempt to provide a model of scientific explanation which, while taking the process of explanation in modern science sketched here as its starting point, may have application to explanation in an even wider context.

CHAPTER 5

THE PRINCIPLE-THEORY-LAW MODEL
OF SCIENTIFIC EXPLANATION

Modern science has two basic tasks: one, the determination of what the facts are, and the other, the explanation of why they are as they are. The determination of the facts is paradigmatically accomplished through the establishment of empirical laws, the expressions of which are seen to provide scientific knowledge, while the explanation of the facts is performed by theories, which provide an understanding of them. The determination of what the facts are presupposes the principle of the uniformity of nature, and their explanation consists in showing how they are but manifestations of the principles of the discipline taken as a whole, and in particular of the contiguity principle of causality. The parameters within which either knowledge or understanding can be said to have been attained are set by the principle of substance.

In the present chapter a model of scientific explanation will be presented in which the conceptual primacy of principles with respect to both laws and theories is demonstrated in detail. On this model – the Principle-Theory-Law (PTL) model – scientific explanation consists essentially in showing by means of a theory how the empirical laws or regularities of a discipline are but the manifestation of the discipline's principles on the empirical level.

1. THEORETICAL MODELS, SOURCE-ANALOGUES
AND ABSTRACTION

Theories explain laws or regularities by depicting states of affairs which can be conceived as underlying them, and which obey the contiguity and other principles of the discipline. The depiction of such a state of affairs thus constitutes a conceptual picture, one whose conformity with reality cannot normally be directly checked when the theory is first put forward. Rather than speak here in terms of a picture however, we shall, for reasons which will become more apparent in the sequel, speak of a model, or more particularly, a *theoretical model*.[1] Thus theories or theoretical models depict states of affairs which it is originally intended are at least *analogous* to what actually obtains in reality.[2]

A theoretical model may or may not depict a state of affairs which is visualisable, or conceivable in terms of everyday notions, its preliminary acceptability depending rather on the extent to which it is in keeping with the principles. Visualisable or everyday states of affairs may be directly relevant to a theoretical model in two ways however, one consisting in its often being the case that such a state of affairs affords the starting point for the construction of the model. When the basic idea for the functioning of a theoretical model stems from such a state of affairs, the latter may be termed the model's *source*. Sources may be actually existing situations or merely imagined ones, but in either case they should themselves appear to embody the principles of the discipline, at least in those respects of central relevance to the explanation of the regularities in question.

-----

[1] As regards the application of this way of thinking to physics, see Hutten (1954), pp. 292–293: "Though physicists do not always speak of a model when considering a whole theory, it is obviously not a serious extension of this term to describe the general frame of concepts underlying a theory in this manner;" and Spector (1965), p. 138: "[I]t is the model which is the heart of the physicist's investigations."

Such models are not models in the sense of the term employed in logic and mathematics. In those disciplines a model is (unintuitively) any state of affairs (mathematical structure) which satisfies a set of axioms and is thus adequately represented by them. Here the term "model" is to have the same sense as it does in the natural and social sciences, such that a model *represents* a state of affairs, and is not the state of affairs that is represented.

[2] In the event that the theory is determined to be essentially correct, one might wish to say that its relation to reality is stronger than one of mere analogy, and that what was originally a model becomes a true description – in this regard see ibid., p. 130. Nevertheless, the term "model" is retained in actual scientific practice, and we can suggest two good reasons for doing so. One is that theories involve idealisation – to be treated below – which means that viewed as descriptions of reality they ought always to be considered false; and the other is that, like models, theories are *applied* to a certain domain with greater or less success, and differ in this way from what are normally associated with the notion of description, namely propositions or sentences which are simply true or false.

The second way in which visualisable or everyday states of affairs may be of relevance to a theoretical model is that they can play an important role in the explanation the theory provides by bringing it down to earth, or connecting it with everyday reality. In this case too the everyday state of affairs should appear to embody the principles of the discipline. In such cases we may speak of scientists' providing an *explanation by analogy to what is familiar*, in which it is shown how the facts or regularities to be explained can come about as a result of the operation of the principles in the same way as they operate in a situation which can be conceived as existing in everyday reality.[3] In this way the state of affairs to which the analogy is drawn functions as a *concrete analogue* to the state of affairs the model represents.[4] In the event that one and the same everyday state of affairs functions as both the source and the concrete analogue it may be termed a *source-analogue*. There is nothing to prevent a concrete analogue better than the source being settled upon after the theoretical model is already constructed however, in which case the source and the analogue would be distinct.

We see then that both theoretical models and their source-analogues may be analogous to the reality the theory is intended to depict; but of the two the relation of greater importance from the point of view of the discipline is that between the model and reality. One reason for this is that explanation in the discipline is ultimately to consist in showing the empirical regularities in question to be the manifestation of the discipline's refined principles, and it is in the theoretical model and not the concrete analogue that these principles are explicitly expressed. Another reason is that the expressions of the regularities cannot be derived from the analogue while they can be from the model, as will be shown below.

One may be considered to have scientifically explained an empiri-

---

[3] In this case, "what the model does is to point to some process which is better known or understood than the relatively unknown process being theorised about." (Graves, 1971, p. 48).

[4] In keeping with Graves (ibid., pp. 50–51), we point out that a number of empiricist philosophers of science conflate theoretical models with concrete analogues, thereby taking the former to be 'merely heuristic devices' which can ultimately be dispensed with. In this regard see e.g. Braithwaite (1953), p. 93; Hempel (1965), pp. 433–447.

cal fact by connecting it to the refined principles via the model, even if no concrete analogue can be found; such an analogue is to be seen rather as a desirable accompaniment of a theoretical model, which can broaden the explanation the model provides by linking it to a common reality outside the discipline. As the exact sciences advance, however, it is becoming ever more difficult to find concrete analogues which even apparently embody the central principles as depicted in various theoretical models.

Theoretical models are as a matter of fact often constructed on the basis of previous attempts at model construction, so that the immediate source of a particular model need not be an imaginable state of affairs. But in order for the explanation the model provides to include analogy to what is familiar, it is sufficient that at some earlier point in the chain of model constructions the link be made to a concrete state of affairs which involves the principles in a form that is still relevant to the theory being constructed, or that such a link can be made directly from the model itself.

Rather generally, we may speak of the model as being an *abstraction* from its ultimate (empirical) source: the model is drawn from the source in such a way that not all of the properties of the source appear in it. This is because only certain aspects of the source are relevant to the explanation of the empirical laws in question, most notably, its causal aspects. But even certain of these may be abstracted from in the model, since the source may not consist of a causally isolated situation, as is presupposed in the expression of the laws. Thus the analogue provided by the model may be termed an *abstract analogue*, as distinct from the concrete analogue which may be provided by the source. To this we may add that a theoretical model may have a number of sources from different areas, some aspect or aspects of each source being included in the model.[5]

The kinetic theory of gases provides a straightforward example of the relation between a model and its source-analogue. Taking the ideal gas model as a theoretical model, we can say that its source is the idea of medium-sized physical objects, e.g. billiard balls, re-

---

[5] The notion of the source of a theoretical model is Harré's, as is the idea that a model can have more than one source: see Harré (1970b), Ch. 2.

bounding upon impact. The model represents gases as consisting of multitudes of volumeless particles endlessly colliding with the walls of their containing vessel in a way similar to that in which billiard balls might collide with the cushions of a billiard table; and the model constitutes an abstraction from its source in that it strips the idea of rebounding physical objects such as billiard balls of such concrete properties as their colour, volume and gravitational attraction to each other and the earth, as well as of the property of losing energy (in the form of heat) upon impact. Further, the analogy provided by the model is that between volumeless particles obeying Newton's laws of motion and gas molecules in a real gas, while the analogy provided by the source is between e.g. billiard balls moving on a billiard table and the molecules of a gas.

2. THE SUBSTANTIAL, FORMAL AND CAUSAL ASPECTS OF A THEORETICAL MODEL

Independently of whether a theoretical model or theory is accompanied by a concrete analogue, the fundamental way in which it explains facts is by showing them to be but manifestations of the principles of the discipline. Since linking the facts to the principles is the main function of a theory, the principles play a central role in how the theory is formed, and may in fact be seen as determining its basic aspects. In modern science the basic aspects of a theoretical model thus correspond to the principles of uniformity, substance and causality, and may be termed the theory's formal, substantial and causal aspects.

The substantial aspect of a theoretical model is its ontology, considered independently of how it operates. The causal aspect consists in those causal relations depicted in the model as operating by contiguity between ontological entities, such that the ontology includes the depiction of causal mechanisms. And the formal aspect consists of the laws according to which such interactions take place: distinct from empirical laws, these may be termed *theoretical laws*.

Like empirical laws however, theoretical laws may be expressed by equations; and while such equations must depict relations between quantities, given the hypothetical nature of theoretical models they need not depict relations between magnitudes, i.e. between *measurable* quantities.[6] Furthermore, in that the principles of primary importance in scientific explanation are causal, the equations expressing theoretical laws are virtually always intended to depict *causal laws*, i.e. regular relations between causes and their effects.

Such theoretical or causal laws must be expressible solely in terms of the notions provided by the refined principles of the discipline, and may themselves directly express such refined principles. Thus, in our example from the kinetic theory of gases, the causal laws according to which the molecules in a hypothetical gas relate to each other and to the walls of the containing vessel are Newton's laws (causal principles) of motion. From this point of view theories may be said to constitute depictions of states of affairs where the only laws in operation are refined principles.

On empiricist philosophies of science the tendency has been to take theories as consisting solely of what is here depicted as the formal aspect of theoretical models, that is, as consisting solely of the expressions of theoretical laws. Thus the deductive-nomological (D-N) model depicts scientific explanation as being by laws, not theories. As an account of actual scientific explanation however, this approach is unable to explain the existence of the model in the context of which the theoretical laws are framed.[7]

------

[6] In this regard see Campbell (1921), pp. 96–97.

[7] Of course it does not suffice simply to say that the equations expressing the theoretical laws themselves constitute the model. As Hutten points out (1954, p. 290): "There are no mathematical models in physics: the equation by itself is not the model. The wave equation is a model only because we know it to represent the spreading of a wave through space." Harré (1970b, pp. 37–38) notes: "[S]ome people still talk of equations as models of motions and processes. At that rate every vehicle for thought would become a model, and a valuable and interesting distinction would be lost." In this general regard cf. also Spector (1965), esp. p. 134, and Graves (1971), p. 53: "Models are ontological hypotheses. It is not clear how, for any given theory, one can determine just what its ontology is, by somehow digging it out of the formalism."

3. THEORETICAL SYSTEMS AND THE DERIVATION
OF EMPIRICAL LAWS

The expressions of empirical laws are specifications of the principle of uniformity arrived at through the application of the principle
of substance to empirical (mensural) reality. Theories or theoretical
models, unlike the expressions of empirical laws, attempt explicitly
to depict the role played by the principle of causality in the state of affairs in which the laws are manifest. Like the expressions of the laws,
however, a theoretical model also depicts a situation constrained by
the principles of uniformity and substance.

As regards the principle of substance, this means that where the empirical scientist creates *mensural* states of affairs which the substance
can neither leave nor enter, with the result that its quantity remains
constant, the theoretical scientist is constrained to create *conceptual*
states of affairs in which the quantity of substance is depicted as being
constant. In physics this consists in holding *energy* constant in the
empirical and theoretical situations respectively. Thus, in accordance
with the substance principle, the experimentalist and the theoretician
each in their own way isolate a substance.

The application of the substance principle on the theoretical level
not only sets fundamental constraints on theory construction, but is a
prerequisite for the application of the theoretical model to empirical
reality. Given the isolation of a substance on both the theoretical and
empirical levels, the theoretician is in the position to claim that the
substance depicted on the theoretical level is the *same* as that isolated
on the empirical level, or that the theoretical substance constitutes the
causally relevant part of the empirical substance – that part of it on
which the regularities it evinces depend.

In either case, in order that reference to the theoretical substance
can be used to provide a causal explanation of empirical regularities,
it must be conceived as including causal relations among its parts, or,
more particularly, as consisting of causal mechanisms. In this way the
substance depicted by the theoretical model may be said to constitute
a *system*, to be understood in the broad sense as a state of affairs
consisting at least partly of regularly operating causes.

The actual application of the theoretical model thus consists in showing how the operation of the causal mechanisms of the theoretical system should give rise to the regularities of the empirical substance in the particular form being investigated. In other words, the empirical laws evinced through operations performed on the actual substance must be shown to be the manifestation of what in the theory are depicted as causal laws (theoretical laws) holding between ontological entities.

Empirical laws are expressed as equations relating the values of mensural properties (magnitudes). In order that they can be shown to emanate from the workings of the causal ontology, the laws according to which those workings transpire must also be expressed as equations, and rules then established for translating values from the one set of equations to the other. This means that all properties in the theory must be expressed quantitatively. Thus where empirical laws are expressed as equations between measurable properties, theories are also in part to be expressed as equations, but between quantified properties of their causal mechanisms.[8] However, due to the possible inaccessibility of theoretical mechanisms, some of the properties represented in the theoretical equations, while being quantities, may not be magnitudes.

In applying the theory, the substance (system) whose properties are represented by the variables in the theoretical laws are identified with the substance the magnitudes of which are related in the empirical laws, or with what is assumed to be its causally active part. Thus all magnitudes represented in the theoretical laws, such as the volume occupied by the theoretical system, and its mass, are identified with the same magnitudes as expressed in the empirical laws. What must then be shown is that, when the magnitudes in the theoretical laws are given the same values as in the empirical law(s) to be explained, the values of the non-measurable quantities, including the substantial and causal properties of the individual causal mecha-

---

[8] This process can also be seen as going in the other direction – cf. Enriques (1906), p. 363: "The hypothesis of an invisible mechanical substratum beneath physical phenomena may be positively interpreted as a *process of association and abstraction* which leads to a representation of the relations of phenomena by means of the *quantitative relations* of certain data, that is, by means of the *equations* that determine the phenomena."

nisms, are essentially the same as the values of particular other magnitudes in the empirical law(s) (cf. e.g. the mean kinetic energy of the molecules of a gas expressed empirically as temperature).[9] Characteristic of such 'other magnitudes' is that, unlike the magnitudes depicted in the theoretical model, their conceptualisation on the empirical level involves notions not found in the refined principles of the discipline.

Theoretical laws obtained for the express purpose of making such a comparison may be termed *derived laws*, an example of such being that of van der Waals. The values they suggest of non-theoretical magnitudes, even when the theoretical model is essentially correct, usually only approximate those actually obtained from instrumental operations. This may be due, for example, to the model's being overly idealised, or to the operation of extraneous causes within the empirically isolated substance. When the respective values obtained are considered to be sufficiently similar – and sometimes this is the case when they are simply of the same order of magnitude – the theoretician can claim to have derived the empirical laws, e.g. Boyle's law, from the theory or theoretical model, and thereby to have explained them.

In this way all empirical properties can be understood in terms of, i.e. be reduced to, theoretical notions, and the form of the empirical law or laws can be seen as resulting from causes operating contiguously through the causal mechanisms depicted in the theory. When the theoretician has succeeded in explaining empirical laws in this way, they too may come to be termed causal laws; and the theoretical model can be said to constitute a detailed picture of what may be supposed to be the causally relevant parts of the substance whose empirical laws have been determined by the experimentalist, thereby warranting the claim that the empirical substance too constitutes or contains a system.

Here we see that the principle of the uniformity of nature plays a central role in the explanation of empirical laws through the specification of the theoretical laws constituting the formal aspect of the

---

[9] In this regard Campbell speaks of a *dictionary* linking the theoretical notions to the empirical ones: cf. above, p. 100, n. 6.

theory, laws from which the derived laws corresponding to empirical laws are obtained. But the principles of causality and substance are also essential, in that it is the causal and substantial aspects of the theory that actually determine the particular form taken by both the theoretical and derived laws.

Key to the connecting of the theoretical and empirical levels however, and thus to the explanation of the latter in terms of the former, is the principle of substance; for it is the substance principle that allows the identification of the theoretical system and some part of the empirical substance, thereby warranting the claim that the theoretical causes are in fact operative on the empirical level. This is not to say that the substance principle provides an algorithm for determining which empirical property or properties are to 'correspond' to the causal properties of the theoretical mechanism; the making of the particular connection depends on the ingenuity of the theoretician. (As history shows, for example, there existed no a priori rule indicating that heat be a form of energy.) As is the case with any principle, the substance principle does not prescribe with regard to the particularities of a situation, but only sets general constraints, leaving to the enterprising scientist the task of filling in the details. Thus where 'correspondence rules' or 'bridge principles' constitute an embarrassing anomaly when the attempt is made to depict theoretical explanation in terms of the D-N model, recognising as it does only a theory's formal aspect, on the PTL model such a connection is warranted by the principle of substance: it is the substance principle that constitutes the bridge linking the reality depicted in the theory to that depicted by empirical laws. At the same time however it must be emphasised that this way of thinking is a concession to that employed in the D-N model, resulting from our here concentrating on the relation between theories and laws. In a broader perspective, if one is to speak in terms of 'bridging' on the PTL model, what is bridged is the gap between principles and laws, and this is accomplished by theories.

In physics the connection between theories and laws is made by taking certain empirical phenomena, such as heat, to be but a form of energy, energy being the substance of the discipline and being explicitly represented in the theoretical model. Thus we obtain the explanation of the empirical level by the notions in terms of which it is conceived being reduced via the theoretical level to categories represented in the refined principles of the discipline, which in turn are specifications of the more fundamental principles expressing categories basic to modern science as a whole: space, time, substance and causality.

Here we see once again the inadequacy if not incorrectness of speaking in terms of theory-ladenness. In the case of empirical notions it would simply be incorrect to say that the notion of temperature, *per se*, is theory-laden. Such a notion does not presuppose the existence of an explanatory theory in order to be meaningful but rather that of certain principles, in particular the principle of the uniformity of nature. And to say of such notions as energy that they are theory-laden would be misleading, in that it would suggest that their source is in a theory or theories, rather than in a particular principle. Calling them principle-laden, on the other hand, draws attention to this fact, as well as to the fact that their dependence on the discipline's principles is a characteristic they share with empirical notions. Here it is to be noted however that the form of dependence is different in the two cases. Where empirical notions presuppose the fundamental principles, theoretical notions are taken directly from the refined principles, except as regards contiguity.

Thus on the PTL model, just as it is essential to distinguish laws from theories, so is it essential to distinguish theories from (sets of) refined principles.[10] Refined principles, such as that of the conserva-

---

[10] The distinction between theories and refined principles is seldom made in the philosophy of science, with scientific work on a discipline's refined principles normally being grouped together with work on what we would here call theories, both being classified under the rubric of theoretical science. As mentioned in Chapter 1 however, Whewell and Poincaré distinguish between theories and refined principles; so does Einstein, who expresses the distinction in terms of constructive theories and theories of principles, citing the kinetic theory of gases as being of the former sort and thermodynamics and the theory of relativity of the latter (1948, p. 54). Einstein also speaks of arriving at the special theory not by reasoning on known facts, but by reflecting on the example provided by the second principle of thermodynamics (1949b, p. 53). Campbell too notes the distinction: "I am not at all sure that the special theory ... is properly termed a theory; I should prefer to call it a principle, and so connect it with the principle of the conservation of energy or the principle of least action." (1923, p. 107n.).

tion of energy, indicate universal features of the particular aspect of reality being investigated by the practitioners of a certain discipline. Theories, on the other hand, are the result of attempts to indicate the nature of those universal features in detail. Thus where each discipline has only one set of refined principles (which in effect define the discipline), due to the empirical inaccessibility of parts of the particular aspect of reality being investigated, as suggested in Chapter 4, the discipline may contain many theories specifying the nature of the reality of interest in different (and incompatible) ways.

4. THEORETICAL MODELS CAN SUGGEST EXPERIMENTS BUT DO NOT DETERMINE THEIR RESULTS

In the event that one theory or form of theory has been generally recognised in a discipline as being essentially correct, its detailed depiction of the causal ontology of the discipline's substance may suggest superior ways of isolating the substance empirically and/or of determining the empirical laws that hold for the substance. For example, as regards the empirical isolation of a substance, advances in atomic theory indicated that what at one point in time were thought to be isolated pure elements in fact contained various isotopes, and led to more stringent methods being used to isolate elements, which in turn resulted in the formulation of simpler empirical laws of proportionality among the elements.

Thus, as has been noted in the previous chapter, theories might well play an important role in the discovery or refinement of empirical laws and the making of such technological advances as the improvement of measuring instruments. Once one has adopted a particular theory, the kinds of measurements one makes, and even the instruments one uses in making them, may change. Nevertheless it must be kept in mind that though a particular theory may determine how an experiment is to be performed, it does not determine the

---

This distinction parallels that made by Agazzi between two kinds of metaphysics: see Agazzi (1988), pp. 11–12, where he distinguishes between metaphysics as the science of those aspects of reality which lie beyond what is empirically ascertainable, and metaphysics as the science of the most universal features of reality. This distinction is of course that between the transcendent and the transcendental.

experiment's results.[11] These results depend rather on the principle of the uniformity of nature, or more particularly on the empirical laws presupposed by the measurements providing the results, as well as on the law manifesting itself in the results.

We today take alterations in temperature, as manifest in thermometer readings, to be the result of changes in the rate of motion of submicroscopic particles. This conception of temperature is based on the kinetic theory of matter, and is in this way theory-dependent. However, thermometers were used to measure temperature long before the kinetic theory was generally accepted; and if the kinetic theory had never been proposed, or another theory had been advanced instead, this would have had no effect on the facts requiring explanation, namely that under certain specified conditions thermometers gave certain particular readings. In the case of temperature then, independently of the mechanism responsible for thermometers working in the way that they do, the principle of the uniformity of nature led scientists to rely on their readings in establishing such empirical laws as the general gas law.

Similar measuring operations performed under similar conditions will always produce similar results. One may of course question whether the results of a particular experiment are correct, but this is essentially an empirical matter, one to be determined by checking whether the experimental operations have been properly performed. Nor need the theory be presupposed in order for the experiment to be meaningful, which is guaranteed by its results constituting a specification of the uniformity principle. Thus it is quite possible to perform an experiment suggested by a particular theory, the results of which imply that the theory is mistaken.[12] What the theory does primarily is provide a particular causal interpretation of the experiment – one which may or may not be in keeping with the experiment's results.

A methodological implication of the above is that it should always be possible in principle to describe an experiment independently of

---

[11] As noted in Dilworth (2007), pp. 137–138.

[12] As expressed by Gingras and Schweber (1986), p. 382: "[T]hough the theory can suggest the construction of a particular apparatus..., this does not mean that the instrument will necessarily yield data consistent with the theory."

any theory in terms of which the results it produces might be explained.[13] However, once a theory explaining those results has been generally accepted, it may often be the case in practice that its notions are employed in describing both the experiment and its results.

5. THE NOMINAL VS. THE REAL ASPECT OF THE SUBJECT

On the PTL model, scientific explanation consists in showing how empirical facts (laws or regularities) are but manifestations of the principles at the core of the science or discipline. This consists in depicting a perhaps hidden reality which both accords with the principles – in particular with that of contiguity – and at the same time gives rise to the facts to be explained. This task is performed by a theoretical model or theory.

As mentioned earlier, a theoretical model may be abstracted from an everyday situation or from earlier modelling attempts, that from which it is drawn being its *source*. In contrast to its source, on the other hand, that to which a theory is applied (the substance of the discipline) may be termed its *domain*, or, when speaking in terms of a theoretical model, the model's *subject*. Thus, according to the PTL model, a theory (theoretical model) is drawn or abstracted from its source and applied to its domain (subject).

In certain cases the source and the domain of a theory may be the same, or may be of the same type. In terms of theoretical models, when this is so the model is a *homeomorph*; otherwise it is a *paramorph*.[14] Thus, for example, many theoretical models in the social sciences are homeomorphs, being abstracted from social situations of the same sort as those to which they are afterwards applied, while the ideal gas model is a paramorph, having as its source the phenomenon of colliding medium-sized physical objects and being applied to gases.

---

[13] In this regard, see Dilworth (2007), p. 140.

[14] This distinction and terminology are Harré's: cf. Harrè (1970b), pp. 40–46. It may be of some interest to note that the PTL model itself is a homeomorph, while the D-N model is a paramorph (being drawn from logic). It may also be noted that the present notion of homeomorph is not that used in mathematics, where one speaks of spaces being homeomorphic if they have the same topology.

The domain of a theory or subject of a theoretical model has two aspects, corresponding to the theoretical and empirical depictions of the substance. One is that which is represented by the theory's ontology, and the other the empirical reality the theory is intended to explain. What is represented by the theoretical ontology is the *real* aspect of the domain (subject), while what is to be explained is the *nominal* aspect. Thus the real aspect contains the causal relations responsible for the nominal aspect's taking the form that it does, a state of affairs which suggests that the real aspect constitutes the (real) *essence* of the nominal aspect.[15] The model represents a reality underlying the phenomena, not with the intention of being a complete description of that reality, but with that of highlighting the essence of the driving forces behind appearances.

Returning to our example from the kinetic theory of gases we may take, as before, the ideal gas model as the theoretical model or theory, and the idea of colliding physical objects such as billiard balls as its source. The theory's domain, on the other hand, consists of gases, the real aspect of the domain being gases conceived of as consisting of multitudes of tiny particles in rapid motion, and the nominal aspect being the known (empirical) gas laws.

6. IDEALISATION

Modern science consists in the application of the principles of uniformity, substance and causality to the world in which we live. This world however is exceedingly complex; in it situations seldom repeat themselves, pure substances are rare, and different causes often act simultaneously. Thus in order to begin to apply the principles, some order must be brought to the world – a procedure described in some detail in Chapter 3. As mentioned there, this ordering involves the creation of ideal states of affairs in which the empirical laws specifying the uniformity principle may be manifest, thereby making the expressions of such laws *idealisations*.

As also mentioned earlier, on the PTL model theories are obtained from their everyday sources by a process of *abstraction*. Taking the

---

[15] A point to be pursued in Chapter 7.

notion of idealisation to be that of conceptually making a state of affairs perfect or ideal in certain respects, we can take abstraction to be the conceiving of a state of affairs in such a way that certain of its original properties are excluded. Thus though abstraction is normally a form of idealisation, it need not be: one can abstract from the colour of a particular object without necessarily conceiving of the object as in any sense more perfect. Nor need idealisation involve abstraction: conceiving of a ball as a perfect sphere does not involve conceptually removing any of its properties.

Speaking rather generally, we can say that where a theoretical model is an abstraction from its source, it is an idealisation of the real aspect of its subject. Thus in our example from the kinetic theory, where the ideal gas model is an abstraction from its source in that it excludes from the consideration of colliding physical objects such properties as their colour and volume, it is an idealisation of the real aspect of its subject in that it depicts gas molecules as moving in completely regular ways. As in the model's being an abstraction from its source, its being an idealisation of the real aspect of its subject is a result of the demand that states of affairs be such that the principles be clearly manifest in them. Newton's principles of motion are not clearly manifest in the movements of billiard balls nor in those of gas molecules, but they are so in the theory abstracted from the former and representing the latter.

Thus, virtually all theoretical models in modern science are idealised abstractions. The scientist constructing such a model has to be careful, however, not to carry idealisation and abstraction too far, for used to excess they can lead to a model that manifests itself in unrealistic ways. Thus, in the present example, by depicting molecules as having no volume in the ideal gas model, one is unable to account for the liquid which results when the substance in question undergoes a change of state.

In recent years there has been a growing literature on the topic of idealisation, and a philosophy of science has been advanced in which idealisation and a subsequent 'concretisation' are seen to be the hallmarks of modern science.[16] While there is much of value in

---

[16] The central work in this regard is Nowak (1980).

this view – including the idea that science is concerned at least as much with essences as with their outward manifestations – it has to date failed to make a clear distinction first between principles, laws and theories, and second between idealisation and abstraction, the result being an overly simple account of idealisation in science.

On the idealisation-concretisation view the scientist begins by attempting (by a process of *abstraction*) to distil the essence of an empirical situation in an idealisational law or an idealised model, and then proceeds to 'concretise' the idealisation by successively adding to it notions corresponding to factual properties in the situation. Two better-known examples where this process is to occur involve Marx' theory of value and the kinetic theory of gases.

As regards the case involving Marx' theory of value we note that the theory or theoretical model is a homeomorph, being both abstracted from and applied to essentially the same state of affairs, viz. economic systems. This aspect of the theory makes it particularly well suited for analysis in terms of idealisation and concretisation. At the core of the theory is the idea that market prices reflect the true value of goods – a notion which can be arrived at by abstraction from real markets; and the fact that the theory is to apply to just such markets makes it possible to effect such an application by successively adding to the theory's core, in an intuitively plausible way, such concepts as correspond to known empirical or 'concrete' factors.

The situation is otherwise however as regards the kinetic theory of gases. Here we note that this theory, considered in terms of the ideal gas model and that of van der Waals, is a paramorph, being abstracted from the idea of colliding physical objects, as well as that what is to be explained, namely the gas laws, are already idealisational in form.

In this context the idealisation-concretisation view faces problems as regards both idealisation and concretisation. Regarding idealisation the difficulty is one of explaining how the various gas models have come to be created by abstracting from situations involving the collision of medium-sized physical objects rather than from situations involving gases. More generally the problem here is one of explaining why there should exist paramorphs in science at all.

The problem regarding concretisation is more subtle. On the idealisation-concretisation view, van der Waals' *law* is to constitute a concretisation of the *law* depicting the behaviour of ideal gases.[17] The idea here is that van der Waals' law, by taking account of the volume of and attractive force between gas molecules, is more factual or concrete than the general gas law. But this conception of concretisation misses completely the role played by van der Waals' *model* in framing van der Waals' law. The question should really be one of whether van der Waals' model is more concrete than the ideal gas model.[18]

In this regard it is to be noted that while van der Waals' model treats of properties not treated in the ideal gas model, these properties are equally as idealised as the others represented in the model; i.e. van der Waals' model, while perhaps less abstract, is no less idealised than the ideal gas model. But is it more concrete? The line of thinking employed here by idealisation-concretisation theorists is that van der Waals' model (law) is more concrete since it takes account of such realistic factors as the volume of the molecules of the gas and the attractive force operating between them. Here we note however that when van der Waals' conception was first advanced the idea that gases consist of molecules was still only an hypothesis, and so the 'concretisation' involved was only a hypothetical one. Thus as regards concretisation in the present example, we have a situation where the result of concretising a particular idealisational law or model is no less idealised than the original, and that, with regard to what it is applied to,[19] it is more concrete than the original only in a hypothetical or theoretical sense.

As indicated by the above examples, the idealisation-concretisation view is overly simple, missing as it does such basic distinctions as that between empirical laws and theoretical models. What is not contested here, on the other hand, is its claim that idealisation plays an important role in modern science – a role

---

[17] Cf. Krajewski (1977), pp. 15–16 and 24–25.

[18] As expressed by Spector: "The very form in which the [van der Waals] equation is stated betrays its origin in the model." (1965, p. 136).

[19] Of course van der Waals' model is more concrete than the ideal gas model with respect to their common source, but this is not the point at issue here.

which, however, can be better understood on the PTL model.[20]

7. EXPLANATION VS. PREDICTION

On the PTL model scientific explanation consists in indicating how particular facts are but the manifestation of the principles of science. We might say that when this has been done one has explained *why* the facts are as they are. And to this it may be added that when the explanation succeeds in meeting the requirements of the contiguity principle in particular, one has also explained *how* the facts have come to be as they are.

In contrast, on the D-N model, explanation consists in indicating how a particular phenomenon is but an instance of a law (conceived of as a true general statement). Thus where on the PTL model of scientific explanation all three of the principles of uniformity, substance and causality are brought into play, with emphasis being laid on that of causality, on the D-N model only the notion of uniformity is considered of relevance. Among other things, this means that where the D-N model treats the notions of explanation and prediction as being essentially the same,[21] the PTL model treats them as being quite distinct.

As is in keeping with the discussion of prediction and technology in Chapter 3, prediction, but not explanation, is on the PTL model particularly aligned with uniformity. To be able to predict how a particular substance will behave under as yet unrealised conditions is to know the laws of the substance with regard to those conditions. Similarly, to be able to *retrodict* how a substance must have been earlier in order to be as it is now is to know the laws that govern its changing. In neither case is any idea required of *why* the substance should behave in this particular way, other than that its doing so is in keeping with the uniformity principle. And no idea is required of *how* the phenomenon results from the operation of causes contiguous with their effects, so long as the rule the phenomenon follows is

---

[20] Another point perhaps of interest is that the idealisation-concretisation approach provides a rather good characterisation of the method employed in the present study!
[21] Concerning which, see Dilworth (2007), p. 6.

known.[22] And the same may be said, *mutatis mutandis*, regarding technology: both prediction and technology are dependent on the principle of the uniformity of nature; and both are essentially empirical.

Just as some phenomena can be predicted without being at that time explainable, so can others be explained without being at that time predictable. In such cases explanation most often consists in indicating the nature of the causal mechanism responsible for bringing the phenomenon about, while the actual preconditions for the mechanism operating in the way that it does are not known, or the particular way in which they influence the functioning of the mechanism is not known.

The distinction drawn between explanation and prediction on the PTL model can be well illustrated in the kinetic theory of gases. To be able to predict how a particular gas will behave under certain conditions is to know the empirical laws governing that gas. This is not to deny, as has been underlined in a broader context, that e.g. knowledge of the mechanism responsible for the gas behaving as it does may provide insight into what the empirical laws regarding the gas actually are. But once these laws are known, it is on the basis of them and nothing else that the behaviour of the gas under various conditions can be predicted.

As regards being able to explain without being able to predict, we might consider once again van der Waals' model and the law derived from it. On the model one could explain[23] the transition of a substance from a gaseous to a liquid state in terms of the attractive force operating between the molecules of the substance at small distances and when moving at relatively low speeds. But the prediction, on the basis of van der Waals' law, of the temperature at which this transi-

---

[22] It is of some interest to cite Nicholas Maxwell in this context: "Failure to articulate the scientific aim of improving understanding may well lead science to degenerate into nothing more that the enterprise of predicting more and more phenomena more and more accurately. Those few scientists who prize the search for understanding above all else, and who protest, will tend to be dismissed as unscientific metaphysicians or philosophers." (1984, p. 99).

[23] Albeit not in a wholly satisfactory manner, since the contiguity requirement is not met.

tion will take place for various substances is quite unreliable.

In the present chapter the PTL model of modern-scientific explanation has been presented, and explicated mainly in terms of the kinetic theory of gases. In the next chapter the model will be applied to a research area in the social sciences, in an attempt to reveal the nature of such research and at the same time indicate the extent to which it may be considered a part of modern science.

CHAPTER 6

THE SOCIAL SCIENCES:
A CONSIDERATION OF ECONOMICS

In Chapters 2 to 4 above the attempt was made to characterise the essential features of modern science; and in Chapter 5 the PTL model of scientific explanation was presented against the background of this characterisation. The three key notions in this approach to science are principles, laws and theories; and the principles concerned have been specified as those of uniformity, substance and causality. These principles, conceived of in their paradigmatic modern-scientific form, imply strict determinism, a perpetually existing physical substance, and the operation solely of efficient causes which are contiguous with their effects.

Actual modern-scientific practice comes closest to its paradigmatic form in the work of such men as Laplace, Faraday and Maxwell in physics, Lavoisier, Dalton and Berzelius in chemistry, and Darwin, Mendel, and Crick and Watson in biology. Since the turn of the twentieth century however, the principles of modern science in its core discipline of physics have come to be qualified in ever more drastic ways, while at the same time the physicalistic methodology suggested by the core principles has been continually extended in such areas as biology and medicine.

In the social sciences, on the other hand, though the mensural methodology of modern science has been widely applied through e.g. the influence of behaviourism, the fundamental principles in their form paradigmatic for modern science have never shown themselves to be as applicable as in the natural sciences. Some of the basic problems regarding their applicability in the social sciences are those of synthesising uniformity and free will, the vagueness apparently inherent in the notion of a social substance, and the dominant position occupied in social thought by the notion of final causes. Here it might be mentioned that a thesis of the present work is that the existence of these problems, coupled with there not being an alternative set of equally clear and coherent principles constituting the core of the social sciences, is responsible for what has been referred to as their inferiority complex, as well as for their attempt to mimic the quantitative methodology of the natural sciences while often lacking an ontological justification for doing so.

It remains an open question however as to the extent to which the view of science developed thus far is applicable to cases where the fundamental principles diverge markedly from the form they paradigmatically assume for modern science. For the sake of interest then we shall in the present chapter attempt to apply our conception of modern science to economics – the social science whose structure is perhaps most similar to that of the natural sciences. In what follows the attempt will therefore be made to analyse mainstream economics (neoclassical microeconomics) generally, and a classic example involving oligopolistic competition more particularly, in terms of notions developed earlier, thereby showing the ways in which economics approaches the paradigm of modern science, as well as the ways in which it diverges.

1. THE PRINCIPLES OF RATIONALITY AND EQUILIBRIUM

Broadly speaking, the science or discipline of economics concerns the human economy, or more particularly the exchange of goods or services between human beings, such exchanges together constituting a market. Exchanges are conceived as being the result of the exercise of the free will of the individuals involved with the intention of maximising the benefits accrued to themselves. In the case of what is termed perfect competition, in which the rate of exchange in the market is not affected by the actions of particular individuals (firms), the unhindered attempts of individuals to maximise benefits to themselves is conceived as constantly moving the market towards a state in which the supply of goods and services is equal to the demand for them. In this way such a market is conceived to constitute a system of rational action tending to a state of equilibrium, i.e. to a state in which supply equals demand. That action on the market be performed with the intention of maximising benefit to the individual actor is termed the principle of rationality;[1] and that the state of the market-system resulting from such action be one of equilibrium is termed the principle of equilibrium.

These principles clearly differ from, while at the same time in certain respects being similar to, the fundamental principles of modern science. What corresponds to the uniformity principle is to be found in the principle of rationality, which depicts a rule according to which economic action is to take place. But this rule allows economic actors to act differently under the same circumstances if they respectively believe themselves to benefit more from different states of affairs (i.e. if they have different preferences).

That which constitutes the substance of economics is the economy, which is similar to the substance of physics in that it involves causation and can constitute a system, but differs from it in that not all practitioners of the discipline agree as to what exactly is to be included in it (the problem of determining externalities). Though the substance of economics differs from that of physics and chemistry in that it is not conceived as existing perpetually, its existence is nevertheless a precondition for the pursuit of the enterprise.

The notion of causality employed in economics is also to be found in the rationality principle, and differs markedly from that of modern science, where it is of an efficient cause. In economics the notion is of a final cause: economic actors employ various means to attain the end of benefit maximisation.[2] The operation of such causes, such that it gives rise to a system in a state of equilibrium, is similar however to that depicted by the modern-scientific principle according to

---

[1] Referred to by Machlup as "the postulate of rational action, the 'economic principle' of aiming at the attainment of a maximum of given ends" (1955, p. 147). That Machlup sees such a principle as being on a par with our (refined) principles of modern science is evident where he speaks of "the fundamental postulates in physics – such as the laws of conservation of energy, of angular momentum, of motion – or ... the fundamental postulates in economics – such as the laws of maximizing utility and profits" (1961, p. 167). On the problems inherent in dealing with the rationality principle from within a Popperian perspective, see Hands (1985).

[2] Cf. Georgescu-Roegen: "Without the concepts of *purposive activity* and *enjoyment of life* we cannot be in the economic world." (1971, p. 282).

which the action of every cause is balanced by an equal and opposite reaction.

An important difference between the principles of modern science and those of economics concerns the way they are received in the two enterprises. In modern science there is a unanimity in the acceptance of the principles and their qualifications as the enterprise develops; in economics this is not the case. Thus, for example, there exists a respected group of economists who argue against the applicability of the unqualified rationality principle to economic situations.[3] A further complication in this regard concerns the role game theory – to which the rationality principle is integral, while competition is not perfect – is to play with respect to the enterprise.[4] In both of these cases the suggestion is being made by particular economists that the basic principles of economics be qualified in such a way that, while they become less simple, they at the same time become more realistic. In a similar vein there is also dissent as regards whether self-interested action in real markets should lead to a state of equilibrium in the relevant sense.[5] But as in the case of modern science we must here too underline the fact that principles function as ideals or paradigms which, while perhaps seldom if ever manifest in a pure form, provide the basic categories of thought for their discipline. This is a role they perform until such time as they are replaced by alternatives of similar simplicity and coherence which are more clearly applicable to the real world, and a new discipline emerges.[6]

---

[3] A leading proponent of this view is Herbert Simon, who suggests employing the notion of 'bounded' rationality, according to which economic actors attempt to 'satisfice' their benefits rather than maximise them, leading to markets in which equilibrium is not achieved. In this regard see e.g. Simon (1979).

[4] See e.g. Johansen (1979), where it is argued that game theory, which prima facie applies best to oligopolistic markets, should constitute the basis of economics.

[5] See e.g. Robinson (1962), pp. 77–80.

[6] In this general regard, cf. Machlup (1956), p. 201: "Such propositions are neither 'true or false' nor empirically meaningless. They cannot be false because what they predicate is predicated about ideal constructs, not about things or events of reality. Yet they are not empirically 'meaningless,' because they are supposed to 'apply' or correspond broadly to experienced events. They cannot be 'falsified' by observed facts, or even be 'proved inapplicable,' because auxiliary assumptions can be brought in to establish correspondence with almost any kind of facts; but they can be

Notwithstanding the raising of such issues, it is clear that the notion of maximising benefit to oneself is central to modern economics, constituting perhaps its most important category, and that at the same time the tendency of economic systems to move towards a state of equilibrium as a result of self-interested action is assumed by the vast majority of researchers in the field. These fundamental presuppositions affect both the empirical and the theoretical aspects of the enterprise, the former of which will be examined in the next section.

2. THE EMPIRICAL FACTS OF ECONOMICS

One can imagine the empirical facts of the natural sciences as consisting of such regularities in the relations between physical objects as could be discovered e.g. by naked-eye observation – much as they were for Aristotle. But for modern science, due to the particular form of the uniformity principle it assumes, the empirical facts consist of regularities (laws) the existence of which can be established by *measurement*.

Turning to economics, one wonders whether a similar transition has occurred, that is, whether for the enterprise in its modern form economic facts are essentially quantitative. That they might be so is supported by the role played by money in economic considerations: to maximise benefit to oneself is conveniently conceived as either getting the most money for a good or service, or giving the least for it, and money is of course quantitative. A perusal of the economic literature suggests however that, while economic facts may often be expressed in quantitative form, this is not always or even regularly the case, and that as a consequence such facts ought not be viewed as being essentially quantitative in the way in which the facts of modern science are. While the data from which the economic facts are culled are almost always quantitative, the expressions of what are taken to be the facts themselves need not be.[7]

---

superseded by other propositions which are in better agreement with these facts without recourse to so many auxiliary assumptions." That economics may be approaching such an intellectual revolution is suggested by the content of Daly (1992).

[7] It is thus in the light of the distinction between data and facts that we must interpret Schumpeter's view that the very subject-matter of economics presents itself in

Similarly, unlike the facts of modern science, economic facts may or may not concern their subject (the economy) independently of time and place, that is, they may either be conceived of as laws holding for all markets of a particular type,[8] or as holding for some particular market. In the case where the economist engaged in empirical research is attempting to determine facts which would obtain in *any* market given essentially the same conditions as obtain in the particular market he or she is studying, we have the element of necessity occasioned by the (economic) facts being but specifications of a presupposed uniformity, viz., that of the principle of rationality.

Further differences are that empirical researchers in economics face particular difficulties which their colleagues in the natural sciences can avoid. Not only are they unable to control the conditions holding in the system they are studying, but such systems are exceedingly complex as compared to physical systems – one aspect of this complexity consisting in the free will human beings are assumed to have.[9] Thus the facts actually ascertained regarding real economies are invariably to some extent provisional, leading some empirical researchers to speak in terms of *stylised facts.*[10]

---

quantitative form (Samuelson, 1972, p. 249), as well as Machlup's comment that "Economics is the only field in which the raw data of experience are already in numerical form." (1961, p. 168).

[8] In this regard see e.g. Schmalensee (1989), p. 959, where he explicitly claims to be seeking economic laws, and Wilhelm Wundt (cited in Machlup, 1960, p. 233), who claims statistical regularities in the social sciences as having the right to be called empirical laws. On p. 1000 of his (1989), Schmalensee draws the same distinction as made in the present study between empirical laws and theoretical models: "[I]nter-industry research in industrial organization should generally be viewed as a search for empirical regularities, not as a set of exercises in structural estimation. [R]esearch in this tradition has indeed uncovered many stable, robust, empirical regularities. Inter-industry research has taught us much about how markets *look*, ... even if it has not shown us exactly how markets *work*."

[9] Here it must be pointed our that there exists a particular tension in the economist's conception of human action. On the one hand the notion of free will is integral to it, since without free will the rationality principle would make no sense. On the other hand, however, no economic actor has the freedom not to follow the rationality principle, which itself determines how he or she is to act.

[10] See Schmalensee (1989) for a use of this terminology, as well as an enumeration of various stylised facts stemming from empirical inter-industry studies.

The discovery or determination of such facts need by no means be a straightforward procedure the results of which are completely uncontroversial. Rather, arriving at facts such as these may be a demanding task involving not only the undertaking of sophisticated and wide-ranging empirical studies, but also a good deal of argumentation for the viability of the methods used and the results obtained.

It is not to our purpose however to debate precisely what the economic facts are considered to be, or what the empirical procedures are via which they are arrived at. Our concern rather is to point out their existence, and in what follows to indicate how theories serve to bridge the gap between them and principles.

3. ECONOMIC MODELS ARE THEORIES

Where principles constitute the most fundamental presuppositions of a discipline, the primary function of theories in modern science is to indicate how a discipline's facts may be conceived of as a natural manifestation of the operation of the discipline's principles in a particular empirical situation. And the scientific explanation of the facts of a discipline consists precisely in linking them to its principles in this way.

Theories in modern science are classically of one of two sorts, being either particle or continuum theories. With regard to modern microeconomics we may say that there are basically three sorts of theory: monopolistic, oligopolistic and perfect competition. Just as in the case of theories of modern science, economic theories may be seen as constituting attempts to explain particular empirical facts by showing how they might come about as a result of the operation of the discipline's principles in the real world.[11] And, as has also been

---

[11] In this regard cf. Robbins (1935), p. 120: "Whether the theory of competition or of monopoly is applicable to a given situation is a matter for inquiry. As in the applications of the broad principles of the natural sciences, so in the application of economic principles we must be careful to enquire concerning the nature of our material. It is not assumed that any of the many possible forms of competitive or monopolistic conditions *must* necessarily always exist. But while it is important to realise how many are the subsidiary assumptions which necessarily arise as our theory becomes more and more complicated, it is equally important to realise how

noted regarding modern science, this is not to say that knowledge of the facts to be explained need precede the formulation of the theory capable of explaining them, nor that theories are incapable of performing other functions besides explaining facts.

Thus, as in modern science, particular facts may warrant explanation, and a particular theory or theories may be advanced with the intention of explaining them. To illustrate this process in economics we shall concentrate on a classic study by Harold Hotelling involving an oligopolistic (duopolistic) model.

On the view advanced in the present work Hotelling's model constitutes an economic theory embodying the principles of rationality and equilibrium, intended to explain a particular economic fact or facts. Hotelling mentions such a fact at the beginning of his study:

This is the fact that of all the purchasers of a commodity, some buy from one seller, some from another, in spite of moderate differences of price. If the purveyor of an article gradually increases his price while his rivals keep theirs fixed, the diminution in volume of his sales will in general take place continuously rather than in the abrupt way which has tacitly been assumed.[12]

In order to explain this fact Hotelling presents a model in which the transportation costs ("a figurative term for a great congeries of qualities") to the customer are included, the result being that he or she will buy one commodity similar to but more expensive than another if its price plus transportation cost is less than that of the alternative. Thus Hotelling, rather than follow his predecessors in conceiving of the market as a dimensionless point, suggests that we:

Consider the following illustration. The buyers of a commodity will be supposed uniformly distributed along a line of length l, which may be Main Street in a town or a transcontinental railroad. At distances a and b respectively from the two ends of this line are the places of business of A and B. Each buyer transports his purchases home at a cost c per unit distance.

---

widely applicable are the main assumptions on which it rests. As we have seen, the chief of them are applicable whenever and wherever the conditions which give rise to economic phenomena are present."

[12] Hotelling (1929), p. 41; next quote, p. 45.

Without effect upon the generality of our conclusions we shall suppose that
the cost of production is zero, and that unit quantity of the commodity is

Market of length $l = 35$. In this example $a = 4$, $b = 1$, $x = 14$, $y = 16$.

consumed in each unit of time in each unit of length of line. The demand is
thus at the extreme of inelasticity. No customer has any preference for either
seller except on the ground of price plus transportation cost. In general there
will be many causes leading particular classes of buyers to prefer one seller
to another, but the ensemble of such considerations is here symbolised by
transportation cost. Denote A's price by p_1, B's by p_2, and let q_1 and q_2 be the
respective quantities sold.

As is clear, Hotelling's model presupposes the principle of rationality: both sellers aim to maximise their own profits, and each buyer
aims to maximise his or her utility (minimise his or her costs). And
the principle of equilibrium is maintained in the market represented
in the model in a way which is in keeping with the empirical facts.
Where on his predecessors' theories a slight reduction in price would
draw all buyers to one seller, leaving the other with a supply for
which there is no demand, on Hotelling's theory there is an equilibrium point flexible with respect to potential changes in transportation
costs, thus providing stability to the system as a whole. In this way
the notions involved in describing the empirical phenomenon of a
gradual diminution in sales of a commodity following a slight rise in
price are *reduced* to the notion of benefit maximisation in a system
which tends to maintain its equilibrium.

4. THE SUBSTANTIAL, CAUSAL AND FORMAL ASPECTS
OF ECONOMIC MODELS

Where the substance of physics is energy, and that of chemistry
is matter, the substance of economics is the economy. Like the
substance of physics, an isolated portion of the substance of economics, viz. a market or markets, is conceived to constitute a system.
Thus markets in a competitive economy are conceived as containing a

causal element (benefit maximising actions) constituting the points of economic interaction between other components of the economy. The basic such components are the goods and services traded; and in its more sophisticated form the market-system includes money, which constitutes an agreement to supply goods or services at any future time. Since money is such an integral part of real modern economic systems, it is virtually always included in economic theorising, leading to the use of notions of sellers and buyers, and prices and costs.

Thus in modern economic theorising the market consists of sellers and buyers of goods and services at particular prices and costs. And the first task of the theorist is to specify the nature of these components of the substance in such a way as can then be integrated in a causal explanation of the relevant empirical fact or facts. As in the case of paradigmatic modern science, this specification of the nature of the substance may be termed the provision of an *ontology*. In our example the ontology consists of two sellers selling a similar good to an indefinite number of buyers located at varying distances from the points of sale. In order to function as a causal explanation this ontology must also include a *causal mechanism* or mechanisms, which in the present case consists in the profit-maximising actions of the sellers, manifest in the system as the respective prices at which the commodities are sold. (It is to be noted that the utility-maximising actions of the buyers are not really part of the causal mechanism in the theory, but constitute rather a precondition for the functioning of the mechanism, and that this is normally the case in modern economic theorising.)

As in the case of modern science, in order actually to constitute a causal mechanism the sellers' actions must fulfil the contiguity principle: there must be a clear connection between an alteration in price and its effect upon sales. This connection is implicit in Hotelling's model, and consists of the medium through which buyers are informed of price changes, which may be e.g. by word of mouth or by advertising.

Thus in the present example of an economic theory we are concerned with a model having both substantial and causal aspects. As is the case with modern-scientific theories, it also has a formal aspect. Like its causal aspect, the formal aspect of Hotelling's model stems from the principle of rationality, which dictates that both sellers and buyers attempt to maximise benefit to themselves. Thus, taking the sellers' locations as the independent variable, the formal aspect of the theory determines the respective prices at which they should sell their goods in order to obtain a maximum profit. This determination is expressed by equations relating profit to prices and transportation costs, and is given a graphical representation by Hotelling in which the prices at which both sellers obtain a maximum of profits constitutes an equilibrium.[13]

5. INTENTIONAL CONSTRUCTS AND EMPATHETIC UNDERSTANDING

As mentioned above, the notion of causality involved in the principle of rationality is teleological, where in modern science the notion employed is of an efficient cause. This, apart from the enterprises' different subject-matters (substances), constitutes perhaps the most fundamental intellectual difference between them, manifesting itself in thoroughly divergent forms of theorising and subsequent explanation.

One way of expressing this difference is to say that where the causal mechanisms of modern-scientific theorising constitute efficient causes, those of modern economics constitute *sufficient reasons*: the actions of economic actors are guided solely by reasons which they consider sufficient for fulfilling the aim of maximising benefit to themselves.[14] Another way of expressing this is to say that economic actors rationally choose the best *means* to achieve this *end*.[15]

---

[13] Ibid., pp. 45–48.

[14] As suggested by Veblen (1909, pp. 176–177), "The distinctive character given to this system of theory by these postulates [principles] and by the point of view resulting from their acceptance may be summed up broadly and concisely in saying that the theory is confined to the ground of sufficient reason instead of proceeding on the ground of efficient cause."

[15] As suggested by Lionel Robbins (1935, p. 125): "The idea of an end, which is fundamental to our conception of the economic, is not possible to define in terms of external behaviour only. [E]ven if we restrict the object of Economics to the expla-

The rationality principle constrains all theorising in economics to explain economic change in terms of rational action, that is, as the result of the deeds of *motivated* beings who have definite *intentions*.[16] Thus the causal mechanism represented in economic theories is a pure economic actor or actors, often conceived as a firm, a causal notion which perhaps might best be characterised as an *intentional construct*.[17]

The difference in the notion of causality employed in economic theorising means that the sort of understanding provided by economic theories differs from that provided by theories paradigmatic for modern science. In the latter case what is required is the conception of a physical state of affairs which may be seen as giving rise to the empirical facts to be understood. In economics, on the other hand, in order to understand the facts in question they must be seen not only as resulting from human acts, but from human acts which seem reasonable given what is believed to be the state of knowledge of the actors.[18] Modern economists would not claim to understand an

---

nation of such observable things as prices, we shall find that in fact it is impossible to explain them unless we invoke elements of a subjective or psychological nature."

[16] "Hence, social phenomena are explained only if they are attributed to definite types of action which are 'understood' in terms of the values motivating those who decide and act. This concern with values – not values which the investigator entertains but values he understands to be effective in guiding the actions which bring about the events he studies – is the crucial difference between the social sciences and the natural sciences," Machlup (1961), p. 352. This of course is not to say that all economic change is intended. In this regard see Weber (1903–06), p. 188.

[17] Cf. Machlup (1967), p. 9: "In this causal connection the firm is only a theoretical link, a mental construct helping to explain how one gets from the cause to the effect." Intentional constructs as depicted here may be seen as constituting a further refinement of Weber's notion of an ideal type: in this regard see e.g. Machlup (1960). Also of interest here may be Freud's depiction of the aims of his psychology: "We seek not merely to describe and to classify phenomena, but to understand them as signs of an interplay of forces in the mind, as a manifestation of purposeful intentions working concurrently or in mutual opposition. ... On our view the phenomena that are perceived must yield in importance to trends which are only hypothetical." (1916–17), pp. 94–95.

[18] "And this imposes on the social sciences a requirement which does not exist in the natural sciences: that all types of action that are used in the abstract models constructed for purposes of analysis be 'understandable' to most of us in the sense that we could conceive of sensible men acting (sometimes at least) in the way postulated by the ideal type in question." Machlup (1955), pp. 151–152.

economic regularity suggested by a model in which the actors were depicted as acting irrationally (i.e. so as not to maximise their benefit), even if the model were useful as a predictive tool. It may thus be further said of economic theories that they attempt to provide an *empathetic understanding*[19] of how certain economic facts have come about as a (possibly unintended) result of particular drives and intentions of human beings, and that the obtaining of such an understanding is precisely what makes sense of the facts, or makes them *intelligible.*[20]

Turning to our example of Hotelling's theory of stability in competition, we see that both sellers and buyers are assumed to be motivated beings acting with the intention of achieving particular ends, the maximising of profit constituting the sufficient reason for action on the part of the sellers, and the acquiring of goods at a minimum of cost on the part of the buyers. Thus the notions of both kinds of economic actor constitute intentional constructs in Hotelling's theory, though the notion of the seller is treated in greater detail. Furthermore, in order to appreciate the explanation the theory is intended to provide, one must be able to imagine oneself acting so as to maximise benefit to oneself, whereupon one is in the position to obtain an empathetic understanding of how a particular market can function such that small changes in price give rise to only small changes in quantity sold.

---

[19] In this regard cf. Adam Smith (1790, p. 9): "As we have no immediate experience of what other men feel, we can form no idea of the manner in which they are affected, but by conceiving what we ourselves should feel in the like situation."

Our notion of empathetic understanding is essentially the same as Weber's notion of *Verstehen*. However, Weber unfortunately seems to see understanding as necessarily being connected with our 'inner experience,' thereby conflating two kinds of understanding and missing the fact that the notion is equally as applicable to the natural sciences as it is to the social sciences: cf. e.g. Weber (1903–06), pp. 64–66 and 123ff.

[20] Cf. L. M. Lachmann (1969), p. 306: "[I]n the study of human action we are able to achieve something which must for ever remain beyond the purview of the natural sciences, viz. to make events *intelligible* by explaining them in terms of the plans which guide action."

6. THE SOURCE AND SUBJECT OF ECONOMIC MODELS

What is the source of theoretical models in economics? In the physical sciences, with their notions of physical substance and efficient causality, the source of theoretical models is most often the idea of states of affairs involving the operation of efficient causes between physical entities. In economics, on the other hand, the substance is the economy, and the notion of causality is teleological. Thus we find that the source of economic models is most frequently the idea of some form of economic activity, and that the source of theoretical models in the social sciences more generally is some instance of social interaction.

As is the case in the physical sciences, a particular social scientific model may have a number of sources, some of which are models constructed earlier in the same discipline. In Hotelling's case, it is clear that his model was influenced in this way by previous duopoly modelling on the part of Cournot, Edgeworth and Amoroso.[21] In this regard it has also been reported that the idea distinguishing Hotelling's model from his predecessors' came to him through a consideration of the behaviour of ice-cream vendors on a beach.

While the source of a theoretical model might also function as an *analogue* aiding a person unacquainted with the model to understand it, it need not do so, and other analogues may be suggested. To allow others better to understand his model, Hotelling draws analogy to the Main Street in a town, or a transcontinental railroad. Thus in the present case we should not speak of a source-analogue, but of a source and analogues which are distinct.

Conceptually distinct from both the source(s) and analogue(s) of a theoretical model is the model's *subject*, that is, that to which the model is applied. In the event that the source and subject are essentially the same state of affairs, the model is a *homeomorph*; otherwise it is a *paramorph*. In economics and social theorising generally theoretical models are most often homeomorphs: they are drawn or abstracted from particular social situations which are the same as those to which they are later applied – or to which they are at least applicable. If we take the main source of Hotelling's model to be the

---

[21] Hotelling (1929), p. 42.

idea of ice-cream vendors on a beach, we should thus consider the model a homeomorph, since it can be applied to just that sort of phenomenon. Hotelling himself claims his model to apply to much more however, suggesting that it can explain such diverse phenomena as why the Republican and Democratic parties differ so little in their politics, why cities have become too large, why Methodist and Presbyterian churches are too much alike, and why cider is too homogeneous.[22]

What a theoretical model is intended to explain constitutes only one aspect of its subject however, namely the *nominal* aspect. Equally as important is the *real* aspect, represented by the theory's ontology and containing the causal mechanism conceived as being primarily responsible for the nominal aspect taking the form that it does. It is with reference to the real aspect of a theory's domain that its nominal aspect is to be explained, the real aspect being represented solely in terms of notions to be found in the principles of the discipline.

In the physical sciences, when a theoretical model is first advanced, the existence of the real aspect of its subject is normally hypothetical. Thus, in the case of the kinetic theory of gases for example, it was not originally known that gases actually consisted of minute particles in rapid motion.

In the social sciences – or more particularly in economics – the existence of the real aspect of a theory's domain is never in doubt. Its causally most relevant part consists of the intentional actions of human beings; and where one is acquainted with the motivations and intentions that guide one's own actions, one has no similar acquaintance with e.g. the subatomic particles of modern physics. On the other hand, however, our knowledge of other people's motives and intentions is quite indirect – a fact providing the main rationale for the school of behaviourism. Thus it may be said that the internal workings of social systems are hypothetical in a sense which at least bears analogy to that in which the functioning of physical systems may be hypothetical.

But more important than this is the fact that, just as in the natural sciences, there is here a categorial difference between what a theory

---

[22] Ibid., pp. 54–57.

represents and what it explains – between the real and the nominal aspects of the theory's domain. Economic actors and their actions as depicted in economic theories occupy a different conceptual level than do the empirical facts explained by such theories. In order to fulfil their explanatory function, economic theories must involve a move to the level of overt behaviour from that of motivations and intentions.[23]

In Hotelling's theory we see this in the implicit understanding that both sellers and buyers are motivated to act with the intention of maximising benefit to themselves. These motivations and intentions are not directly accessible to observation. The step is made from this hypothetical, explanatory level to the empirical level to be explained through the notion of *prices*. Thus it is through a consideration of overtly manifest prices, against the background of the principle of rationality, that one can obtain an idea of the causal mechanism determining the form taken by a particular market.

7. ABSTRACTION AND IDEALISATION

Broadly speaking, scientific theories may be seen as being abstractions with respect to their sources and idealisations with respect to the real aspect of their domains. Many of the properties present in the source are not represented in the theory; and the real aspect of the domain is depicted in the theory as being of a simpler nature than it actually is in reality. Both abstraction and idealisation in theorising are connected to the fact that theories depict situations in which the principles of the discipline are clearly manifest.

As regards economic theorising generally, the most notable source of idealisation is the principle of rationality, which suggests that all economic actors should act perfectly rationally, i.e. so as to promote

---

[23] This categorial difference is evidenced e.g. as regards the potential testability of hypotheses from the different levels. As expressed by Machlup: "Can there be any doubt that a direct empirical test of the motivations behind businessmen's actions, such as a test whether their decisions are made in an attempt to maximize profits, would be 'more difficult,' to say the least, than a test that higher prices are paid for bicycles?" (1956, p. 202).

their self-interest.[24] As mentioned earlier in this chapter, a number of economists have reacted to the dictates of this principle as being overly idealised.

Turning to our example we can see particular instances in which the principles of rationality and equilibrium have led to abstraction and idealisation. Taking the source of Hotelling's theory to be a situation involving ice-cream vendors on a beach, not only is abstraction made from the physical constitution of the beach and its surroundings, but what is a two-dimensional area in the source is reduced to a one-dimensional line in the theory. Hotelling's model is drawn from its source in such a way as to include only those of the source's aspects considered to be particularly characteristic of it, while at the same time providing a conceptual state of affairs which is in keeping with the principles of rationality and equilibrium.

Similarly, the market as depicted in the theory is idealised as compared to the situations the theory is intended to represent. Hotelling mentions a number of these idealisations himself: the cost of producing the commodity is taken to be zero; consumption is temporally and spatially uniform; and there is no qualitative difference in the commodities offered by the different sellers. To this we may add that in the model the various costs incurred by the sellers are not included, such as the cost of gaining information regarding the location of potential customers and the behaviour of the other seller, and the cost of evaluating various pricing strategies. Here we see that the idealisation involved is also directly connected to the fact that what is being emphasised in the model is a representation of a situation in which the relevant principles (viz. of rationality and equilibrium) are clearly manifest.

---

[24] In this context reference may be made again to Weber's notion of an ideal type. In his words: "Such ideal-typical constructions are exemplified by the concepts and 'laws' [principles] formulated in pure economic theory. They state what course human conduct of a particular kind *would* take *if* it were *strictly* rational (in a subjective sense built into the type), unaffected by error and emotion and, furthermore, *if* it were completely and uniquely oriented toward only one objective, namely, economizing. In reality, action rarely corresponds to the idealized conduct hypothesized in the ideal type . . . and even then it does so at best approximately." *Wirtschaft und Gesellschaft*, Tübingen: J. C. B. Mohr, 1922, p. 4; quoted in Machlup (1960), p. 238.

The considerations of the present chapter have shown that in the main the depiction of modern science in terms of principles, laws (facts) and theories is also applicable to the social science of economics, though the nature of its principles differs from those paradigmatic for modern science. In the next chapter we shall apply our conception of modern science to the issue of natural kinds in the attempt to depict how it is to be treated from a modern-scientific point of view.

CHAPTER 7

NATURAL KINDS

The world as we experience it is not a constantly changing mosaic of sensations in which no order can be discerned, but one which evinces a particular regularity from day to day and year to year. For modern science this regularity rests on the principle of the uniformity of nature, which states that natural change is lawful or takes place according to rules. In conjunction with the other two principles fundamental to modern science, it claims that changes in the world are the result of causes contiguous with their effects operating in a regular fashion on a perpetually existing substance. In this way modern science attempts to explain change in terms of non-change: perpetually existing (types of) causes operate on a perpetually existing substance in a regular manner.

In this world of lawful change some aspects appear to change more slowly than others, or appear not to change at all. It thus becomes part of the task of modern science to explain why change occurs more rapidly in certain regards rather than in others, or, conversely, to explain why certain kinds of entities tend to persist, as well as to clarify the relations among such kinds of entities.

1. WHAT ARE NATURAL KINDS FOR MODERN SCIENCE?

What we perceive to be natural kinds – such as the different sorts of chemical elements and the various species of animals and plants – are, from the point of view of modern science, *forms of substance*. In this way the principle of substance is central to the modern-scientific conception of natural kinds. What distinguishes the various forms are the laws that hold of them. Thus gold is a natural kind since certain laws hold of it, for example that it melts at a particular temperature; and it is a natural kind different from that of silver since the laws that hold of it (the facts regarding it) are different from those that hold of silver.[1] The situation is similar as regards living species: tigers are distinguished from lions according to what can be said of the species *tiger*, as regards both tigers' appearance and behaviour – according to whether the same biological facts do or do not apply to them.[2] Due to the relative complexity of life-forms however, as well as their evolutionary nature and the volitional aspect we assume certain of them to have, their properties and behaviour are not so regular as are those of entities in the inanimate world, and the criteria by which they are judged to be or not to be of the same kind are not so straightforward.

Here it is to be noted that what in modern science is taken to be a natural kind is a form of substance, not particular instances in which the form is manifest. Thus gold is a natural kind while a particular lump of gold is not; and the species *tiger* is a natural kind, while an individual tiger or group of tigers is not. So, just as science with regard to change is concerned with laws and not their instantiations *per se*, science with regard to non-change is concerned with natural kinds and not their instantiations *per se*.

Another point that may be noted here, but to which we shall re-turn below, is that in modern science it need not be the case that all classificatory systems are intended to divide the world into natural kinds. Generally speaking, the nature of such a system will largely depend on its purpose, and there may be other purposes than making groupings according to form of substance.

If we consider biology independently of the way it is pursued in modern science, we could say that its substance is life quite gen-erally – that the subject-matter of the discipline of biology is life. For modern science, however, substance in its paradigmatic form is

---

[1] Cf. below, p. 162, n. 24 and the accompanying text. Note that, as is in keeping with the principle of substance, solid gold and liquid gold are not conceived to be different kinds, but to be different states of the same kind. In this regard, cf. Locke (1690), III. vi. 13.

[2] As expressed by Kitcher with respect to biological species, "statements ascribing to members of a species appropriately chosen properties would be candidates for laws about the species" (1984, p. 313). Cf. also Bigelow et al. (1992, p. 380): "[I]t makes no sense to speak of a natural kind, e.g. being an electron, independently of the laws which govern its behaviour." David Hull's treatment of this question (1981, p. 141) is vitiated by his failure to distinguish between laws and principles, and his consequent ascription of an explanatory function to the former.

physical, which has the effect on biology as a modern-scientific discipline that its interest is only in physical forms of life (and not, say, supernatural ones).

As intimated in introducing this chapter, what are generally regarded as instances of either physico-chemical or biological natural kinds tend to persist; but this is by no means a requirement for their constituting natural kinds (forms of substance) when seen from the viewpoint of modern science. Nor need modern-scientific natural kinds exist in nature; they might just as well be created in a laboratory. The requirement for something's constituting a natural kind is rather that laws can be established regarding it (including its changes of state); and what is further sought is that it be a form of substance which is not decomposable into other forms of substance, i.e. that it not consist of natural kinds which are different from it and from each other. Thus, for example, bronze can be decomposed into copper and tin, which makes the latter more fundamental from the point of view of modern science, and strengthens their case for natural-kind status.

Another role decomposition plays with regard to modern-scientific natural kinds is that it can give rise to a hierarchy of kinds. This is most clearly the case where the ingredients constituting a decomposed kind differ from it but have the same form as each other, such as when gold is decomposed into atoms of gold. In modern science, with its conception of substance as physical, this (ontological) hierarchy goes from physics, whose forms of substance include atoms and subatomic particles, to chemistry, whose forms of substance are the chemical elements, to e.g. geology, whose forms of substance are kinds of rock. In this way too scientific natural kinds from different ontological levels constitute the subject-matter (substance) of different scientific disciplines.

Thus we see rather generally that it is not the relative simplicity of the laws that hold for particular forms of substance that determines whether they constitute more or less fundamental natural kinds from the point of view of modern science, but the extent to which they can be decomposed.[3]

---

[3] A question taken up by van Brakel (1986, pp. 306–307).

2. NOMINAL AND REAL ESSENCES: KEY TO THE
UNDERSTANDING OF NATURAL KINDS

On the modern-scientific conception of reality, the empirical laws holding for various forms of substance are conceived as stemming from regular causal relations in which the causes are contiguous with their effects. In cases where the contiguous nature of these relations is not evident, theories are propounded involving hypothetical ontologies in which the contiguity requirement is met. According to the PTL model, what the theoretical ontology represents constitutes the real aspect of the theory's domain, while what the theory is intended to explain constitutes its nominal aspect.

Applying this reasoning to forms of substance conceived as natural kinds, we should say that theoretical ontologies represent the *real essence* of natural kinds, while at the same time being intended to explain their *nominal essence*. Thus where the nominal essence of a scientific natural kind consists of the empirically (mensurally) manifest regularities in its nature and behaviour, its real essence consists of the causal ontology considered responsible for the nominal essence being as it is.[4]

The basis of this way of viewing natural kinds can be found in the work of John Locke.[5] Differences between his view and that of modern science, or the way his view has been refined in modern science, include the idea that the real essence of things, such as may consist of their atomic structure, need not be or remain unknown to the in-

---

[4] In this way we can understand Kitts' and Kitts' claim that "it is not the discovery of some common property, ostensive or otherwise, which leads us to suppose that species have essences. The search for essences is prompted by theoretical necessity." (1979, p. 621), as well as Wilkerson's saying "It is precisely because gold has a certain atomic number that it has certain properties (its being malleable, fusible, etc.); it is precisely because an oak has a certain genetic constitution that it has certain properties (a characteristic way of growing and reproducing itself), and so on." (1988, p. 29; see also p. 41).

[5] See Locke (1690, III. iii. 15), where real essence is depicted as "the real internal, but generally (in substances) unknown constitution of things, whereon their discoverable properties depend." The nominal essence of a thing, on the other hand, is a complex abstract idea of the thing, the idea the name of the thing stands for. Thus: "the nominal essence of gold is that complex idea the word gold stands for, let it be, for instance, a body yellow, of a certain weight, malleable, fusible, and fixed." (ibid., III. vi. 2).

quiring scientist,[6] and that the nominal essence of a form of substance is not an abstract idea but consists of properties paradigmatically determined by operations of measurement.[7]

Thus the nominal essence of gold consists of those mensural laws that hold of it, including e.g. those regarding its melting point, specific gravity and conductivity. The real essence of gold, on the other hand, consists in its atomic constitution, as is determined by the number of protons in nuclei of atoms of gold, and the number and position of the electrons orbiting each nucleus.[8]

In this way we see that there is a clear difference of level between the nominal and real essences of a scientific natural kind. A number of the properties that can be attributed to the nominal essence, such as melting point, specific gravity and conductivity in our present example, cannot meaningfully be attributed to the real essence. This leads us to say that the real essence of a particular natural kind itself

---

[6] As expressed by Mackie (1976, p. 78): "Since we can equate Locke's real essences with what we should now call the molecular and atomic structure of things, we may say that many real essences that were unknown in Locke's day are now pretty thoroughly known by chemists and physicists."

[7] Cf. ibid., p. 85: "Locke's main purpose would have been better served if he had identified the nominal essence rather with the set of characteristics of which the complex idea in question is the idea.... This way of putting it still secures the point behind Locke's talk about an idea, namely that these characteristics count as the nominal essence because we know them and use them as criteria of recognition, we associate the name 'gold' with the conjunction of them: it is a human mental operation that groups these characteristics and no others together and uses them in classification. By contrast the real essence of gold is the real internal constitution which all pieces of gold have, and on which all these defining characteristics in fact depend, but which we may well know little or nothing about, though we surmise that there is something of the sort."

[8] Virtually the same distinction as that presented here has been made by F. A. Paneth between what he terms simple and basic substances: see his (1931), pp. 150ff. In this regard see also Mackie (1976), pp. 91, 96: "If we have framed and confirmed a theory about the atomic structure of what we now recognize as gold, and then consider ... the counterfactual possibility that something with this same internal constitution was ... not shining yellow in colour, not malleable, not fusible, not soluble in *aqua regia*, and so on, ... we would express this by saying that *gold* might not be yellow, etc., whereas if we contemplate the counterfactual possibility that something with a different internal constitution had all these features, we would say not that (some) gold might have a different internal constitution, but only that something else might look and behave like gold."

constitutes a natural kind, one different from and on a different level than that of the natural kind whose real essence it is. Thus, as suggested earlier, atoms of gold also constitute a natural kind, one lying on a level different from that of gold.

The nature of these different levels is difficult to characterise. We depicted them above as constituting an *ontological* hierarchy. But an *epistemological* ordering is also involved, one which goes in a direction opposite to the ontological. So the difference of level may be termed either ontological or epistemological, depending on the context. Thus, when knowledge is not had of the real level, e.g. when the theory depicting it is still hypothetical, the difference of level may be considered epistemological. When, on the other hand, the kinds of existents are being considered, e.g. gold as compared with atoms of gold, the difference may be considered ontological.

As may be seen from the above discussion, difference of level is relative. Where atoms of gold constitute the real essence of gold, their characteristics may constitute the nominal essence of a yet deeper level, consisting, say, of elementary particles such as quarks and leptons. In this case, then, a particular formation of elementary particles may be considered to constitute the real essence of atoms of gold.

While the notion of the relativity of the nominal and the real was not recognised by Locke, present considerations suggest, in keeping with Locke, that for modern science the real essence of natural kinds lies in their internal constitution as versus their external relations. This both paradigmatically is the case and historically has been the case not only in modern physics and chemistry but also in modern biology.[9]

3. NATURAL KINDS IN BIOLOGY

Where the substance of physics is energy, and that of chemistry is matter, the substance of biology is life. Thus where natural kinds in physics and chemistry are forms of energy and matter respectively, natural kinds in biology are forms of life.

---

[9] Cf. Wilkerson (1988, p. 41): "Whereas the internal or microscopic properties directly determine the superficial or macroscopic, the converse is just not true."

The difficulties involved in fitting the subject-matter of biology into the conceptual mould demanded by the principle of substance constitute one of the primary reasons for biology's not being paradigmatic as a modern science, as are physics and perhaps chemistry. While the existence of life is a prerequisite for that of biology, life itself cannot be quantified, as can the substances of physics and chemistry; furthermore, where the appearance or disappearance of energy is difficult to conceive of, our minds experience no great hindrance in thinking of life as not having been, or as ceasing to be. (The thought of its coming to be, however, is more difficult.)

The fundamental principle of modern biology is the principle of evolution, which suggests particular refined principles of substance, uniformity and causality for biology. As regards substance, the principle of evolution states that life (physically) evolves, which implies that the forms of the substance of biology, i.e. biological natural kinds, while relatively constant for periods of time, nevertheless change. As regards uniformity, the principle states that this change takes place in such a way that the kinds whose manifestations are better adapted to their environments continue to be manifest so long as the relevant environments do not change; and as regards causality that the continued existence of the kinds whose manifestations are better adapted is due to their being better adapted.

As is the case regarding the refined principles of the physical sciences, the principle of evolution calls for specification on both the empirical and theoretical, or nominal and real, levels. As regards the nominal level the task is one of specifying the nature and behaviour of the various forms of life, both past and present. As regards the real level the task is that of explaining the nominal level by indicating in detail both the causal mechanism or mechanisms responsible for the relative constancy of natural biological kinds as well as that or those responsible for their changing.[10] Historically speaking, the specification of the nominal level of biology has been pursued with success though intermittently at least since the time of Aristotle,

---

[10] A distinction similar to that made here between the principle of evolution and the empirical and theoretical aspects of biology has been made by Wassermann (1981, p. 416), where he views the so-called theory of evolution "as a *hypertheory* which explains classifiable evolutionary phenomena in terms of subordinate classifiable theories of 'evolution-specific mechanisms.'"

while the greatest strides on the real level have been made only since the determination of the structure of DNA in the 1950s.

The evolutionary nature of the natural kinds of biology makes their treatment more complicated than that of natural kinds in the purely physical sciences. The smallest units of biological natural kinds are species, followed in increasing size by genera, tribes, families and so on up to the most inclusive biological kinds, viz. the animal, plant and protist kingdoms. Note that what is being spoken of here are biological natural kinds, which may or may not be captured in a biological classificatory scheme or taxonomy, as mentioned above and to be returned to below.

Biological natural kinds, according to modern biology, are ontologically completely determined by the principle of evolution, such that the kind common to two biological entities is fundamentally dependent upon their ancestral trees.[11] The natural kind *species* differs from the other natural biological kinds however, one of its primary distinguishing marks being, in the case of sexually reproducing organisms, that a criterion for two of them (of the appropriate genders) being of the same species is that they be capable of producing fertile offspring. This quality thereby constitutes a central aspect of the nominal essence of biological species, at least as regards many of the 'higher' life-forms. Species identification within biology is still in many cases problematic, however; and it is in any case to be noted that the above depiction functions at best as indicating a *criterion*, and not as indicating either necessary or sufficient conditions for something's constituting a species – as will also be returned to below.

When knowledge of the breeding capabilities or ancestral trees of two organisms is lacking, they may be (perhaps incorrectly) taken

---

[11] Cf. Darwin: "[T]he characters which naturalists consider as showing true affinity between any two or more species, are those which have been inherited from a common parent, and, in so far, all true classification is genealogical; . . . community of descent is the hidden bond which naturalists have been unconsciously seeking, and not some unknown plan of creation, or the enunciation of general propositions, and the mere putting together and separating objects more or less alike. . . .

"If it could be proved that the Hottentot had descended from the Negro, I think he would be classed under the Negro group, however much he might differ in colour and other important characters from negroes." (1859, pp. 404, 407).

either to be or not to be of the same kind on the basis of other overt characteristics, such as their manifest physical structure or *phenotype*. With regard to the first division of levels of biological entities, an organism's phenotype is of the nominal level, while its *genotype* is of the real. In other words, the form of life embodied by a particular organism is determined by its genetic constitution.[12] This is not to say that the environment of an organism's ancestors has played no causal role in the determination of the form of life instantiated by the organism, but that for modern biology the immediate cause of the organism's taking the form it does – i.e. the causal mechanism responsible for the manifestation of that form – is the genetic constitution of the organism. Thus in the case of any particular organism, a change in its environment need not lead to a change in its phenotype, whereas a change in its genetic constitution will do so.[13] Even when modern biologists consider how particular phenotypes have evolved due to changing environments, they conceive of that evolution of phenotype as being mediated by a genetic change.

A further difference between biological and physical natural kinds that deserves mention is that in the paradigmatic case of physico-chemical natural kinds, i.e. the chemical elements, the real essence of the kind itself constitutes the whole of the kind as manifest in the nominal essence. Gold consists of nothing but atoms of gold. In the case of biological natural kinds the real essence does not constitute the whole of what is manifest in the nominal essence: organisms do not consist solely of their genetic constitutions. Here we have a situation in which the real essence of the form of substance in question constitutes only the causally contiguous aspect, i.e. the causal mechanism, directly responsible for the kind's taking the particular

---

[12] "Thus, despite superficial appearances to the contrary, it is supposed that the members of a species have some underlying trait which serves to distinguish them from the members of other species and, most significantly, serves as the foundation for an explanation of their exclusive relationship to their species. Since the discovery of the structure of genetic material it has been possible to get at this underlying trait not only through the manifest properties and the reproductive behavior of organisms, but more directly by means of chemical techniques." (Kitts & Kitts, 1979, p. 622).

[13] Here we intend that such a change not merely be one in what is termed the 'junk DNA' of the organism.

form it does. The situation is similar to that in the economic example considered in the previous chapter, where the substance consisted of the whole economy, while the causal mechanism consisted only of the intentional actions of the sellers. Such intentional actions, represented by intentional constructs, constitute the real essences of particular social situations as characterised in modern economics.

4. ON IDENTIFYING NATURAL KINDS

Viewing the difference of level between the nominal and the real as being epistemological, one can consider the situation in which knowledge is had of the nominal essence of a natural kind while the nature of the real essence is unknown. In other words, we might consider the case where the empirical laws regarding a form of substance are known, while the theory depicting the underlying causal mechanism responsible for the laws is either lacking or hypothetical.

One way of characterising the nominal essence of the kind in such a situation is as those of its properties that serve to distinguish instances of the kind from those of other kinds.[14] Here not only is the epistemological rather than the ontological aspect of natural kinds emphasised, but so is the notion of difference rather than similarity. Thus when the modern scientist is concerned to identify a particular entity lacking knowledge of its real essence, he or she does so by a process of elimination in which the properties of the entity are compared with those of established kinds. If the known properties of the entity are not sufficient to allow its kind to be uniquely determined, it must be examined more closely, perhaps in the form of an experiment, until through a better knowledge of its properties it can be determined to be of but one particular kind. Similarly, if at first the entity appears to have the properties of a particular kind, it may nevertheless turn out that an experimental investigation of it would serve to show it not to be of that kind – as one can imagine a particular piece of fool's gold (pyrite) being distinguished from real gold,

---

[14] In this regard, cf. Harré (1970b), p. 300: "Some core of its totality of qualities, as manifested, serves as the *nominal essence* of a thing, that is as that set of its manifested qualities which are required to remain unchanged for it to be reidentified as the same thing."

say, on the basis of its having a different melting point. And in the event that the given form of substance can be distinguished from all known kinds, it may be considered to constitute a new kind.

Here however modern science, like Aristotelian science, makes a distinction between accidents and properties, where only the latter are related to the essence of a kind of thing. It is an entity's *properties* that are manifestations of the nominal essence of its kind, even if certain *accidents* invariably accompany instances of the kind. Thus, for example, the colour normally associated with gold would be an accident with respect to a particular piece of gold if it were found to be due to the presence of copper in the sample. And we might say that paradigmatically for modern science the properties of a kind are quantitative, being more particularly magnitudes in the case of the nominal essence of the kind.

For Aristotle however, as for Hume and all succeeding generations of empiricists, the sort of distinction made in modern science between different ontological levels is lacking, or is at least less developed. So even on those empiricist views where essences are admitted, the relation between the (real) essence of a kind of thing and the kind's manifest properties does not involve the same sort of transcendence of level as it does for Locke and modern science.[15] Thus for modern science the properties of a thing as can constitute the nominal essence of its kind are manifestations of the kind's real essence, while its accidents are not. So knowledge of the real essence of the kind may show some of the thing's distinguishing marks, such

---

[15] This seeing of reality as consisting of different levels makes of modern science what John Blackmore (1982, 1983) calls an *in*direct philosophy or epistemology. While this characteristic is not shared by Aristotelian science, it is by Platonic science, where the world of ideas lies on a deeper level than does the world of sense. One might want to argue that there is also a difference of level in Aristotle, the (final) cause (essence) lying at a deeper level than its effect or effects (properties of the thing). Nevertheless, there is an important sense in which the Platonic and physicalistically-based modern-scientific philosophies or worldviews are similar to one another and different from the Aristotelian. First, the former both question the viability of sense-impressions as a source of information regarding the deeper-level ontology and in this way are anti-empiricist in their orientation; and second, their deeper-level ontologies are conceived of as constituting whole hidden worlds consisting of entities existing in and of themselves which are in some way more real than the entities populating the world perceived by the senses.

as the yellowness of gold, to be merely accidental, i.e. not to be manifestations of the real essence, such that the thing would not be of a different kind if it lacked them. But it is only when the nature of the real essence of a kind is known that this relation can be of help in identifying things of the kind.

On the Aristotelian view the essence of a thing determining its kind is given by its (true) definition; in modern science on the other hand, rather than being determined by *definitions* of substances, kinds are determined by *operations* on substances, operations the meaningfulness of which presupposes certain ontological principles. Definitions and the correct use of language generally are not central aspects of the enterprise of modern science.[16] The determination of whether a thing is of a particular kind, when knowledge of its real essence is lacking, is paradigmatically the result of an active process involving the treatment of the thing with instruments in a highly artificial, idealised situation, in order to determine its behaviour under various conditions. Thus we see, among other things, that the specification of both the nominal and real essences of a kind paradigmatically involves idealisation.

Also, rather than be an either/or affair as is suggested by the application of formal logic to taxonomic thinking, particular entities may turn out to constitute borderline cases between different kinds. More generally we may say that, unlike much empiricist thinking, modern science does not employ notions of necessary and/or sufficient con-

---

[16] Cf. Whewell (1847, Part 2, p. 14): "It is absolutely necessary to every advance in our knowledge, that those by whom such advances are made should possess clearly the conceptions which they employ: but it is by no means necessary that they should unfold these conceptions in the words of a formal Definition." Cf. also Campbell (1920), pp. 52–53, as cited in Chapter 1: "If we boldly refuse to pay any attention to logical canons our difficulties vanish at once. Our words then are not instruments by means of which the process of thought is conducted, but merely convenient means of recalling to our minds thoughts which have once passed through them or of calling up in the minds of others thoughts which are passing through our own. [Our objector] raised his difficulty first by asking for a definition. We should have refused to give one. No student of science has ever felt the smallest need for a formal definition of silver; our words are perfectly effective in calling up the thoughts we desire without one, and in admitting the right of anyone to ask for one we were encouraging a very dangerous delusion."

ditions in its original identification of natural kinds, but the notions of *paradigm* and *criteria*.[17]

When the nature of the real essence of a kind is not known, whether or not an entity is to be considered as being of the kind is determined by comparing the entity to a paradigm. Such paradigms, or *types*,[18] constitute 'best examples' of the kind – instances of the kind which the researcher believes on the basis of his or her scientific intuition to embody most clearly the nominal essence of the kind, i.e. to constitute the clearest manifestation of the kind's real essence.[19] In this way paradigms become standards of compari-

---

[17] In this regard see Boyd (1991), pp. 142–143: "Thus some paradigmatic cases of natural kinds . . . are counterexamples to the claim that . . . natural kinds must be defined by necessary and sufficient conditions. I conclude that the requirement that natural kinds have such definitions is to be diagnosed as a holdover from traditional empiricist linguistic precision."

[18] Cf. Whewell (1847), Part 1, p. 494: "*Natural Groups given by Type not by Definition.* [T]hough in a Natural Group of objects a definition can no longer be of any use as a regulative principle, classes are not, therefore, left quite loose, without any certain standard or guide. The class is steadily fixed, though not precisely limited; it is given, though not circumscribed; it is determined, not by a boundary line without, but by a central point within; not by what it strictly excludes, but by what it eminently includes; by an example, not by a precept; in short, instead of Definition we have a *Type* for our director.

"A Type is an example of any class, for instance, a species of a genus, which is considered as eminently possessing the characters of the class. All the species which have a greater affinity with this Type-species than with any others, form the genus, and are ranged about it, deviating from it in various directions and different degrees."

As expressed by Ruse (1976, p. 251): "[I]n his recognition that the biologist must work with types and not necessary and sufficient conditions Whewell's thought here, as so often elsewhere, contains seeds of the directions to be taken by modern scientists." Note that this notion of type differs from that of Linnaeus (cf. Stearn, 1971, pp. 246, 247). Many who argue against 'essentialism' in science conceive of essences in Aristotelian terms, i.e. in terms of definitions providing necessary and sufficient conditions; an example is Sober (1980, pp. 379–381).

[19] In this regard, see Caplan (1981), p. 136: "Genotypes are the 'hidden' substrates that allow the grouping of organisms into sets of creatures. [T]he point of attending to similarity of phenotype, behavior, chemistry, or any other organic property is to facilitate inferences about the genetic factors that produce these properties." And he continues: "One might disagree about the utility or possibility or validity of looking to genotypic similarity as a definition of species identity, but, this does seem to be what biologists do."

son,[20] similar to but lacking the conventional aspect of the units constituting standards of measurement (treated in Chapter 3); and things are classified on the basis of whether they can be assimilated to a particular paradigm. Such paradigms need not even be instances of the kind, but can be idealised constructs[21] developed through an acquaintance with various instances of the kind.

Thus, on the view adopted in modern science, the manifestation of the nominal essence of a kind of thing – a kind as conceived in the absence of knowledge of its real essence – does not constitute sufficient conditions for the thing's being of the relevant kind. Some entity of a different kind, as determined by the kind's differing real essence, may manifest itself in all known ways as do things of the kind in question; or certain traits embodied in the paradigm, such as the yellowness of gold, may be accidents and not stem from the instantiated real essence. In other words, it is not until scientists believe themselves to know the real essence of the kind of thing in question that they can feel certain of the thing's being of a particular kind. Nevertheless, due to the role played by paradigms in the conceptualisation of reality, the presence of paradigm-based features of the nominal essence does not constitute merely empirical evidence for the presence of an instance of the kind, but rather *criterial* evidence for its presence. At the same time, however, identification by assimilation to a paradigm leaves open the possibility of there existing borderline cases, so the grouping of things in this manner need be neither exhaustive nor exclusive.

Scientific certainty as regards the identification of a particular entity, i.e. the determination of its kind, may be considered to be obtained only when knowledge is had of the kind's real essence.[22]

---

[20] Cf. Wittgenstein's characterisation of a paradigm as "something with which comparison is made" (1953, § 50); in this regard see also Kripke (1972), p. 122.

[21] In this way they can resemble Weber's *ideal types*, as referred to in the previous chapter. If one wishes to trace them back in the history of ideas, the connection can undoubtedly be made to Plato's notions of *Form* or *Idea*, and *paradigm* (*paradeigma*), the latter of which is a transcendent model embracing all the Forms.

[22] Thus: "The trait in virtue of which the type is selected is typicality, not of phenotype, but, of genotype. . . . Often it is unclear how representative any organism is of the population it is being used to exemplify. But uncertainty over the adequacy of evidence presented by a given specimen does not show that illustrating a typical genotype cannot be the valid function of type-specimens. It only shows that this

When the scientist knows the real essence of a thing's kind and the way in which the real essence manifests itself empirically, he or she is furthermore in the position to distinguish the entity's true properties from its constantly accompanying accidents, and thereby form a clearer idea of what the nominal essence of its kind actually is. In this way the efforts of the theoretical scientist can lead to alterations in a natural-kind taxonomy, such that empirically similar sorts of individuals are classified differently – the classic example of whales' not being fish resting, from the modern-scientific point of view, ultimately on known differences in their evolved genotypes. In a similar way knowledge of the real essences of particular sorts of individuals with disparate manifest properties may lead to their being re-classified as being of the same kind, such as one can imagine having been the case with respect to dry-land and wet-land primulae.[23]

As conceived in modern science, the real essence of a form of substance determines the laws that hold for that form of substance and thereby what kind of substance it is. In this way the notion of natural kind is a complement to that of empirical law, representing the type of substance for which a particular uniformity holds.[24] Thus where the principle of uniformity is central to the notion of empirical law, the principle of substance is central to that of natural kind, as noted at the beginning of this chapter.

---

representativeness is hard to ascertain." Caplan (1981), p. 137. In this regard cf. Locke (1690), IV. vi. 4: "No proposition can be certainly known to be true, where the real Essence of each Species mentioned is not known." And ibid., III. vi. 50: "For if we know not the real essence of gold, it is impossible we should know what parcel of matter has that essence, and so whether *it* be true gold or no."

[23] An inorganic example is given by Harré (1970b, pp. 198–199): "Diamond, black carbon and graphite manifest different characteristics, but they are all carbon because they are alike in the electronic structure of their atoms, that is they have identical constitutions, and belong to the same natural kind."

[24] "Thus, whenever we speak of 'silver' or 'iron,' we are implying that certain laws are true, namely the laws asserting the association of the properties of silver and iron. If very high electrical conductivity was not associated with a brilliant white colour and solubility in nitric acid to give a solution in which ammonia forms a precipitate soluble in excess – and so on – we should not speak of silver; and if strong paramagnetism was not associated with the power of combining with carbon to form alloys which can be tempered, we should not speak of iron." (Campbell, 1920, p. 43).

5. SETS, CLASSES, INDIVIDUALS AND NATURAL KINDS

Natural kinds are forms of substance, and in a way similar to that in which empirical laws constitute the uniformity principle, entities of a particular kind constitute that kind. Furthermore, similarly to the way in which the uniformity principle does not consist of the laws that constitute it, neither does a natural kind consist of the entities of the kind. Kinds in modern science have a conceptual priority with respect to the entities constituting them.

Though the conception of natural kinds being advanced here is intended to apply to natural kinds as they are conceived of in modern science, if one gives it a broader application it may throw light on the present debate concerning natural kinds in the philosophical literature. Against the background of the present view then, it would appear that much confusion has arisen in the discussion due to a failure to distinguish between three related entities. These are: natural kinds as forms of substance; the groups of entities constituting particular natural kinds; and groupings determined by particular taxonomies. This confusion has undoubtedly been abetted by the presently dominant formalist trend in philosophy which favours the application of the mathematical notion of a set or class in the present context.[25] Thus, on this way of viewing natural kinds, rather than be a particular form of substance, the kind *gold* is nothing other than the set consisting of all the gold in the universe, and similarly the kind *tiger* is nothing other than the set of all tigers. Furthermore, particularly in the latter instance, it is assumed that the natural kind in question is to be delineated by whatever taxonomy that treats of things of that kind, i.e. that all (scientific) taxonomies should delineate natural kinds.

Despite the severe problems this view has given rise to, particularly with regard to the notion of species, virtually everyone contributing to the current debate shares it. The reaction of a number of them to these problems has been to claim that species are not (as a matter of fact!) natural kinds but something else – most commonly,

---

[25] For an influential contribution to this confusion, see Quine (1969), p. 118: "Kinds can be seen as sets determined by their members."

individuals[26] – rather than question whether natural kinds should be conceived of as classes or sets.

By conceiving of natural kinds as being forms of substance which may or may not be delineated by particular taxonomies, on the other hand, these problems do not arise; and the notion of natural kind is more easily applied to the sorts of things, such as chemical elements and biological species, which are normally considered to be instances of natural kinds.

The conceiving of natural kinds as forms of substance does not eliminate the notion of class from the discussion however, but admits it with respect to the *manifestations* of natural kinds. Thus, while biological kinds are not classes, biologists nevertheless *classify* individual living entities, i.e. they conceptually place them in classes; and such classes may be intended to reflect differences of natural kind. However they need not be so intended. For example, in cladistic classification in biology divisions are based on the sequence of evolutionary branching, whereas in classical biological classification divisions are based on the grade of organisation of the entities being classified. Cladistic classification is thus more in keeping with distinctions of natural kind as understood against the background of the principle of evolution (whether or not this is the intention behind the cladistic system). But whether cladistic classification is to be preferred to classical depends on the purpose to which the classification is to be put, part of which purpose is undoubtedly to make communication easier between scientists. Thus not only need actual scientific classification not reflect differences of natural kind, but it might very well not be intended to do so, and its not doing so may be an advantage.

Furthermore, even classification according to natural kind is highly misleading if it is taken to indicate an exclusive and exhaustive ordering of the kinds themselves. The manifestations of some natural kinds may fit into no class in a taxonomy consisting of exclusive and exhaustive classes, while the manifestations of others

---

[26] Arguments for species being 'individuals' may be found e.g. in Dupré (1981) and Hull (1981); an argument for their being classes (sets) may be found e.g. in Caplan (1981). The general (mis)identification of kinds with classes or sets is explicit in Ruse: "Ghiselin and Hull argue that species are not natural kinds at all: They are not classes with members." (1987, p. 230).

may fit into more than one. In either case the problem lies with the nature of the taxonomy, and not with the notion of natural kind *per se*.[27] On the other hand, by admitting a classificatory scheme based on paradigm-thinking rather than on what we have elsewhere termed 'box-thinking,'[28] these kinds of problems could be avoided.

Another striking way in which this formalistic mode of reasoning manifests itself in contemporary discussion is in the conceiving of essences in terms of necessary and/or sufficient conditions.[29] Given this way of thinking, it is an easy step to the claim that e.g. species do not have essences, and thence to the conclusion either that species are not natural kinds or that natural kinds do not have essences. By taking the (nominal or real) essence to be a *paradigmatic form*, on the other hand, natural kinds themselves become essences, and entities can be considered as being of a certain kind even if they are not, or do not contain, perfect embodiments of that form.

A further point that might be mentioned regarding the current debate – one related to the above – is that from a philosophical point of view which does not wish to prejudge the empiricism/realism issue, the question regarding natural kinds is fundamentally a conceptual and not an ontological or factual one. The broader philosophical question is not whether e.g. species *really* are natural kinds, or *actually* have essences, but rather how we are to conceive of species.[30] Similarly the question is not whether, if something were to behave exactly like water but have a different molecular structure, it would still *be* water, but how we are to conceive of the chemical elements and their compounds. The general question in the present case concerns how we are to conceive of the 'groupings' or kinds found in nature, and any reasoning which might lead to such a conclusion as

---

[27] Just as, we might say, the problem of induction is a problem for formal logic, not for modern science.

[28] In Dilworth (1992), pp. 207–210.

[29] As referred to on p. 161, n. 19, with reference to Sober. In this regard see also Mellor (1977), esp. pp. 306, 309–310; and Dupré (1981), esp. pp. 88–89.

[30] Thus the philosophical discussion is not one "of the ontological status of species" (Caplan, 1981, p. 136), but of their conceptual status. This widespread misconception may also be found e.g. in Sober, where he treats it as a question of fact "[w]hether species are natural kinds or spatio-temporally extended individuals." (1980, p. 360).

that there are no natural kinds simply begs this question.[31] By conceiving of natural kinds as forms of substance, on the other hand, both chemical elements and biological species and other forms of life can be treated as natural kinds. Such a treatment is more in keeping with both the scientific and common-sense views of the situation, allowing the former to explain the latter, and at the same time leaving open the possibility of noting important differences between sorts of natural kind.

6. ON DIFFERENCE OF LEVEL AND THE EPISTEMOLOGICAL STATUS OF ATTRIBUTIONS OF NOMINAL AND REAL ESSENCE

Our original dividing up of the world into natural kinds is not arbitrary. As animals living on this planet we must make certain distinctions in order to survive, and be able to recognise things of the same sort when we see them a second time. We learn to distinguish kinds on the basis of differences in their manifest properties before perhaps going on later to define them verbally in terms of those properties. We do not determine what the word "gold" is to refer to ('fix its reference') by verbally defining it, but by learning to distinguish the particular form of substance that is gold from other forms[32] – a task for which modern science is particularly well suited.

Thus scientists distinguish gold from non-gold on the basis of its having certain properties which non-gold, including fool's gold, does not have. These properties are manifest under various conditions, a number of which are brought about artificially, such as is the case e.g. in determining the melting point of the substance. And it is in fact through the performance of various operations on matter that the form of substance is isolated and a paradigm established.

Now, having isolated gold in this way, i.e. through determining its nominal essence, certain things may be said about it, such as that it

---

[31] As expressed by Aristotle: "We must attempt to recognise the natural groups, following the indications afforded by the instincts of mankind, which led them for instance to form the class of Birds and the class of Fishes." *Parts of Animals*, Book I, $643^{b}10$–12.

[32] On this point, and others of relevance to the present chapter, see the discussion in Whewell (1847) Part 1, pp. 469–542.

has a particular melting point. How should such a statement be construed from a philosophical point of view? Here we should say that its status depends on the state of knowledge of those either uttering it or hearing it. The determination of the melting point of gold, such a fact constituting part of the nominal essence of the (form of) substance, alters the concept of gold. And we should thus say that, at least for those who are aware of this alteration, the statement that gold has this particular melting point is analytic.[33]

Thus the concept of gold is here associated with its nominal essence. Should it also be associated with its real essence, i.e. should the statement that gold has atomic number 79 also be viewed as analytic?

In seeking to understand why the nominal essence of gold is as it is, the theoretical scientist postulates that gold has a particular constitution, itself not directly manifest, which is responsible for that nominal essence. The experimental investigation of this postulated constitution often takes the form of a refined analysis of the nominal essence of the substance, and may well lead to the constitution's being re-characterised. Thus, for example, it was by bombarding gold film with alpha particles that Rutherford firmly established the atomic nature of gold, while at the same time contributing to the characterisation of that nature by showing atoms of gold to have minute massive centres – their nuclei. It was apparent that this constitution of gold, these atoms, did not themselves have the properties (nominal essence) of gold, but quite different ones. Thus, in effect, a new natural kind had been discovered, one lying on an ontological/epistemological level different from that of gold.

From the point of view of the present analysis we should thus say that the having of atomic number 79 is part of the nominal essence of this new kind, while also being part of the real essence of gold. In this case then it would be analytic to attribute the atomic number 79 to atoms of gold, but not to gold itself. This would set a limit on what is to count as the concept of gold – a limit determined by

---

[33] In this regard cf. Wittgenstein (1953), § 79: "The fluctuation of scientific definitions: what to-day counts as an observed concomitant of a phenomenon will tomorrow be used to define it." For Campbell's views on this issue, see his (1920), pp. 45–55. See also Kant (1783), Ak. 267.

gold's nominal essence – while at the same time recognising the intimate relation between gold and its atomic number.

But what is the nature of this relation? It is not causal in the prima facie sense of the principle of causality, for there the notion of causality is tied to that of change, and here there is no change. Furthermore, here the relation is between different levels, and it would seem that the straightforward use of the notion of causality involves taking cause and effect as being on the same level. The present case seems to concern the same state of affairs seen from two perspectives, the one 'macro' and the other 'micro.' But the situation is not symmetric: atoms of gold are not dependent on gold for their existence, while the opposite is the case.

Here, following Graves,[34] we shall call the relation one of *grounding*, and take it as generally depicting the connection between different ontological/epistemological levels, so as to include the relation between theoretical ontologies and empirical laws. Nature being as it is, the real essence of gold is constrained to manifest itself as the nominal essence of gold. This we take to be the case independently of whether we know what the real essence is, or whether we are familiar with all properties attributable to the nominal essence.

Scientists characterise the nominal essence of a kind of thing in terms of those of the thing's properties which serve to distinguish it from other kinds of thing. These are the criteria for picking out things of the kind. If we take the concept of a particular kind of thing to correspond to its nominal essence, then statements characterising attributions of nominally essential properties to the thing are analytically true. And correct statements delineating the real essence of a substance, such as that gold has atomic number 79, may be considered true by virtue of the relation of grounding.

In this way theories can provide *taxonomic principles* for the classification of kinds. At the same time, however, it is to be kept in mind that theories are typically originally hypothetical, and so the grounding they originally provide is so as well; and a particular conception taken to be that of a real essence at one time may be replaced with another such conception with a change of theory. Thus taxo-

---

[34] Cf. Graves (1971), p. 45. Crompton (1992, pp. 146, 147) also uses the term "grounding" in essentially this way.

nomic principles likewise, while intended to be grounded in reality, may be revised with change of theory.[35] But we may say rather generally that where ontological principles determine the *categories* of a particular science or scientific discipline, well-established theories, through their provision of taxonomic principles, determine the *kinds* (of entities) with which the discipline is concerned, i.e. determine the discipline's *ontology*.

The principle of substance is central to considerations regarding natural kinds, natural kinds being forms of substance for modern science. In the next chapter, where we shall consider the topic of probability and confirmation, the principle of the uniformity of nature plays a central role.

---

[35] Thus, as expressed by Harré, a taxonomic principle "should be treated as a proposition immune from falsification for the time being" (1970b, p. 216). Here of course we are assuming such a principle to be one for classification according to natural kind.

CHAPTER 8

PROBABILITY AND CONFIRMATION

Various theories of probability have been propounded in modern times. In the present chapter we shall consider the most important of them with regard to their applicability to modern science, and suggest a particular view – in which the nominal and real aspects of reality are distinguished – to be the superior alternative. We shall also consider the application of probabilistic and similar forms of thinking to questions regarding the respective acceptability of theories, laws and principles.

1. GENERAL CONSIDERATIONS REGARDING PROBABILITY

In considering the nature of probability, and its possible uses in and applications to modern science, it is to be kept in mind that a fundamental distinction exists between what may be called the qualitative and quantitative notions of probability. The qualitative notion of probability may be associated with the simple adverbial and predicative expressions "probably" and "is probable," while the quantitative notion involves the assignment of a *measure*. While the qualitative notion may be reformulated in quantitative terms, namely as a probability greater than zero (is possible) or 0.5 (is probable), its normal employment does not commit the user to being able to specify the quantitative nature of the relation between the evidential basis of the probability claim and the claim itself. With the use of the quantitative notion, on the other hand, the user may be viewed as being so committed.

Quite generally, we can say that a probability is invariably a probability of something's being the case. A similar view has been expressed to the effect that probabilities are paradigmatically, if not always, probabilities of (the occurrence of) *events*. But this latter view seems too narrow, and does not do justice to a variety of instances in which there is an agreement in our intuitions as to the viability of employing probability notions, especially qualitative ones. Thus, for example, we should have little hesitation in applying the notion of probability to such a case as that regarding the size of quasars, saying that quasars are probably of the order of physical magnitude of galaxies, considering the evidence we have for the great quantities of energy they release; and here we are not speaking of an event, at least not in any straightforward sense, but of something's being the case.

Another delimitation of probability notions which seems too strict is the suggestion that they are to have relation only to *future* events or states of affairs. But can one not say, for example, that Socrates probably knew of Democritus; or, given adequate geological evidence, that the probability of a particular volcano's having erupted in a certain year was such-and-such? Similarly, one should be able to apply probability notions to certain present states of affairs, such as that involving quasars mentioned above, of which one does not have complete knowledge. In fact, it is the notion of complete or sufficient knowledge that is key in this regard; and, as will be suggested below, in the case of modern science it is precisely in those cases where such knowledge is lacking that the concept of probability finds its application.

A third delimitation of probability notions that does not seem entirely warranted is the denial of their applicability to statements, hypotheses, conjectures, guesses and other locutions which we would normally say are either true or false. For such locutions describe states of affairs, and, given that what we mean by a statement's being probable or having a particular probability is that it is probably true or that the probability of its being true is such-and-such, then we are only expressing in different terms the probability of what the statement describes being the case. And this is quite in keeping with what we expect of probability notions.

2. TWO SENSES OF THE WORD "CONFIRM"

By allowing that we can speak of the probability of statements or hypotheses (being true), we bring the notion of probability closer to that of *confirmation*, for we can also apply the latter notion to such locutions. But here we must distinguish two senses of the word "confirm," one being 'to corroborate, or add support to (a statement, etc.),' and the other, 'to make certain, verify, put beyond doubt.'[1] It is clear that if we are to align probability with confirmation, we must take "confirm" in the first sense, which does not imply certainty. If one were to use this term with regard to the empirical laws of modern science, however, as has been demonstrated in Chapter 3, due to their intimate relation to the uniformity principle it is the second sense of the term that is to be employed.

3. EVIDENTIAL BASIS VS. SUBJECT-MATTER

In considering the nature of probability claims it is important to distinguish the evidential basis of such a claim from its subject-matter, or what it is about. The former often manifests itself in the form of particular data, of which we have knowledge. The latter, on the other hand, consists of the state of affairs (those states of affairs) regarding which knowledge is lacking, and about which we are using the data to gain information.

This distinction is of particular importance with regard to the statistical or frequency interpretation of probability. It is one thing to say that (quantitative) probability is nothing other than what can be obtained on the basis of an empirical investigation of relative frequencies, and another to say that what a probability claim made on such a basis is *about* is itself relative frequencies. This distinction has been overlooked not only by critics of the statistical interpretation, but also by its supporters.[2] There is nothing in principle to prevent one's taking the evidential basis of a probability claim to be the relative frequency of the occurrence of events in an empirical sample, and its subject-matter to be a unique state of affairs.

---

[1] Both senses taken from the *Oxford English Dictionary*.

[2] Even von Mises misses it, where he is led to say: "The phrase 'probability of death,' when it refers to a single person, has no meaning at all for us." (*Probability, Statistics and Truth*, W. Hodge, 1939, p. 15; cited in Kneale, 1949, p. 165). For an exception, see Georgescu-Roegen: "The point, as I have argued in opposition to Mises, is that every prediction is about a single event." (1971, p. 172).

4. METHODOLOGICAL GROUNDS AND INDUCTIVE PROBABILITY

As suggested in Section 1 of this chapter, the making of acceptable quantitative probability determinations requires there being a specifiable quantitative link between the evidence for the determination and the value determined. The nature of this link is dictated by the methodological foundation of the probability claim, or the *rules* in accordance with which the claim is made. These 'methodological grounds' tell us first what is to *count* as evidence, and second how to obtain a probability value on the basis of that evidence – a procedure which includes a weighting of the evidence, i.e. a quantitative delimitation of the (discrete or continuous) field of evidential possibilities together with rules indicating what place any possible evidence is to be assigned in the field. All probability determinations are based on evidence and require methodological grounds, whether those grounds be implicit or explicit.

It is thus with some scepticism that we regard Carnap's suggestion that what he calls inductive or logical probability is on a par with, yet distinct from, statistical probability, and occurs in contexts of a different kind. According to Carnap, a statement of inductive probability is to be such that, if an hypothesis and evidence are given, the probability can be determined by logical analysis and mathematical calculation.[3] This requirement, however, in that it asserts that there must be a clearly delimitable relation between the evidence for a probability claim and the value asserted by that claim, is simply the demand that (part of) the methodological grounds of such claims be specifiable. Though there is nothing to prevent Carnap's having developed his view so as to provide new methodological grounds for probability determinations, what he here terms 'inductive probability' is but an aspect of any acceptable determination of quantitative probability.

---

[3] Carnap (1955), p. 271.

5. SUBJECTIVE PROBABILITY AND THE IMPLICATIONS
OF A PROBABILITY CLAIM

In Section 3 above, two aspects of probability claims were discussed: their evidential basis and their subject-matter. Another aspect is what they *imply*, or, in certain contexts, what they *mean*.[4] If a person says that it will probably rain tomorrow, the evidential basis of their claim may be today's weather report, while its subject-matter is tomorrow's weather. But the claim might *imply,* among other things, something about their state of mind, which in turn might set limits on what may be considered rational behaviour on their part. Thus their claim might be taken to indicate that they believe that it will rain tomorrow; and, assuming they believe this, that it would be foolish of them to begin preparations for a day at the beach.

The notion of what probability claims imply is of particular relevance to the subjective interpretation of probability, according to which probability determinations represent *degrees of belief.* Now if it is intended by subjectivists that probability is nothing other than degree of belief, then their view is quite unacceptable: not only is there no way of directly quantifying the 'degree' of a person's belief, but different people can have different beliefs on the same evidence, and there would be no sense in speaking of certain of those beliefs as more justified than others. Unlike in the case of objective knowledge, there would be a probability particular to each person, with no guarantee that it would be the same for everyone.

Another possibility is that one take one or more persons' degree(s) of belief as the *evidential basis* of a probability claim. But even if people's beliefs could function as evidence for actual probabilities, it would only be natural to ask what the basis was of their beliefs. Once this basis was determined, it would then constitute direct evidence for the probability of the state of affairs in question, evidence which could moreover be used to judge the viability of the beliefs.

A third line might be to say that what probability claims are *about* are degrees of belief. But, as is suggested by the above example, this is only to confuse the subject-matter of a probability claim with its

---

[4] For a distinction between the grounds and the meaning (sense) of a probability statement, see Toulmin (1958), p. 76, and Harré (1970b), pp. 160ff.

implications. While one's saying that it will probably rain tomorrow might be construed as implying that one believes to a degree greater than 0.5 that it will rain tomorrow, the claim is not about the person's belief, but, as mentioned above, about tomorrow's weather.

The place of the notion of degrees of belief in probability theory, as suggested above, is at best that of their being *implied* by probability claims. This place is rather peripheral however, and as a consequence we may conclude that the notion of subjective probability is of but little relevance to the theory of probability.

6. KNOWLEDGE-RELATIVITY AND THE PROPENSITY INTERPRETATION

Another interpretation of probability is the propensity or 'objective' view, according to which probability is a property of a state of affairs. With regard to the term "objective" as used here, we note that it is (or should be) intended in the sense of 'residing in the object,' and not in the sense that would suggest this approach to be necessarily more objective than other approaches when it comes to the method by which probability assignments are made.

An important implication of the propensity view is that probabilities, in being objective properties of states of affairs, are not knowledge-dependent. Thus the probability of a particular die's turning up an ace being 1/6 is a property of the situation in which the die is cast, and is not dependent on our knowledge of that situation. But is this so? Let us say that the die is cast and turns up a five. This does not refute the objectivist's thesis. But let the die be cast again in a situation that is qualitatively *identical* to the former one. According to common sense and the principle of the uniformity of nature, since the situation is qualitatively exactly the same, the die will again turn up a five. The reason we speak of probabilities in the case of dice throwing is that we do not *know* precisely what the situation is in which a die is cast.[5] The probability of obtaining an ace is not a

---

[5] As regards the application of this way of thinking to quantum mechanics. cf. d'Abro (1939, vol. 2, p. 957), where he asks: "What requirements must be realized for us to be justified in asserting that the conditions under which successive experiments are performed are identical? Until a rigorous method of determining the iden-

property of the situation, but a function of our knowledge of the situation; and, following Laplace, we can here claim more generally that for modern science probability is to be understood as being knowledge-relative.[6]

The propensity interpretation can have counter-productive effects if applied to science, for it tends to condone ignorance.[7] If the probability of something's happening is simply taken to be a property of the state of affairs in which it happens, then once one has probabilistic information there is no impetus towards a more detailed investigation of the state of affairs. One can, with equanimity, say that that is just the way things are, and no compulsion need be felt to attempt to discover more rigorous laws according to which the events in question occur.

7. NOMINAL VS. REAL PROBABILITY DETERMINATIONS

When a particular probability distribution is empirically determined always to obtain given conditions of a certain kind, then, on the

---

tity of such conditions is given, we shall always be free to suppose that differences in the observed results arise precisely because the conditions are not identical."

[6] As expressed by Campbell (1920, p. 159): "Probability is a measure of our degree of knowledge concerning the happening of an event." Similarly, Democritus saw chance as a regularly acting cause which is obscure to us; in this regard cf. Guthrie (1965), p. 419: "Chance as a subjective notion can take its place in the system without prejudice to the ruling idea of an all-pervading necessity." One here sees how the propensity interpretation is positivistic or empiricist in its orientation, since it takes limitations on the acquiring of empirical knowledge as being a direct indication of a characteristic of the world being investigated. Here, as elsewhere, the ontological collapses into the epistemological on the empiricist view.

[7] Regarding the relevance of this claim to quantum mechanics, see Blackmore (1983), p. 29: "What Einstein and Planck wanted was a theory of matter and light *as they existed independently of observation and sensory measurement* which would make the complementarity principle and the inexactness (not uncertainty) principle more understandable. But Bohr and Heisenberg, by denying that there was a 'meaningful' physical world beyond possible conscious observation and experiment, denied the possibility of greater understanding than the inexactness and complementarity principles could offer. In short, for Planck and Einstein the Copenhagen interpretation and its phenomenalist restriction on further inquiry seemed to stand in the way of a more complete understanding of micro-reality."

modern-scientific conception and as suggested by Laplace,[8] the distribution is considered not to be due to hazard, but to a regularly operating cause. Were there not such a cause in operation, we should not understand why just that probability distribution is manifest given conditions of that kind.[9]

In some cases the cause may be relatively evident, as when our concern is with the casting of dice; in others it may be hidden. When it is hidden, scientists may *theorise* about its nature. In the best of cases they can construct an acceptable theory which enables them empirically to recognise the difference in the conditions under which each of the elements in the distribution is manifest, thus allowing them to replace probability with certainty. But even if their theory is unable to do this, it might nevertheless explain why the distribution as a whole takes the form that it does. Thus, for example, while on the empirical level one can see that a certain proportion of the children of parents with blue eyes and brown eyes themselves have blue eyes, by moving to the level of the genetic theory scientists consider themselves to be able to understand this phenomenon through an appreciation of the nature of its cause, or real essence. And, given that they are warranted in claiming that their knowledge has been extended to the nature and functioning of genes, we might go so far as to say that the probability value obtained from an empirical investigation of a (necessarily) finite number of cases is not quite the correct one, and furthermore what the value should be if based on an indefinitely large sample.

Here, then, even assuming the probability value obtained to be the same in the two cases, the evidential basis of the latter determination would differ from that of the former. While evidence for the empirical determination would also be evidence for (or against) the theoretical, if it is claimed that the theoretical mechanism exists, evidence directly affirming its existence over and above its ability to produce probability values near to the empirical ones would also

---

[8] P. 393 in Harré (1967). It is of some interest to note that not only Democritus, but other early thinkers such as Hippocrates, and later philosophers such as Spinoza, conceived of chance as but another name for ignorance.

[9] As suggested by Campbell, the results of such trials "are not due to pure chance because they are not distributed at random." (1920, p. 213).

constitute evidence for the viability of the values it suggests. Also, the methodological grounds on which the theoretical probability assignment is based, in that they concern how the actions of the theoretical mechanism should be manifest empirically, would also differ from the rules employed in the purely empirical case.

This brings us to a consideration of current usage of the terms "a priori" and "a posteriori" in probability theory. The distinction normally drawn can with justification also be made here, where probability determinations derived from theory have their ultimate basis in the (a priori) principles of the discipline or science in question, while determinations of the empirical sort are a posteriori. This distinction can also be applied to simpler cases. Thus we should say that an unloaded die, cast under normal conditions, has the a priori probability 1/6 of turning up an ace, or that the a priori probability of getting zero on a fair roulette wheel is 1/37 (or 1/38), and contrast this with the a posteriori probability of such occurrences based on observed frequencies, the value of which may be different.[10] Modern science thus has to do both with a priori and a posteriori probability, which respectively correspond to its investigation of the real and nominal aspects of reality.

8. METHODOLOGICAL REQUIREMENTS OF PROBABILITY LOCUTIONS

In order to make a quantitative probability determination, whether a posteriori or a priori, not only must one be able to put the methodological grounds of the determination in a quantitative form, but one must also have a clear concept of its subject-matter. For example, if one says that the probability of a particular coin's turning up heads on the next flip is one-half, one must have a clear notion of what it means for the coin to turn up heads.

This point is not problematic in games of chance, where the occurrences with regard to which probability calculations may be made are clearly delimited. But when we move over to considering states of affairs in a broader context, this clarity may be lacking. Thus, for

---

[10] For a discussion of this point, see Kneale (1949), § 37.

example, one cannot speak of the probability of the cat's being on the mat, if one is not clear as to whether for this to be the case the cat's tail must also be on the mat. In other words, the application to the cat of the predicate 'being on the mat' must obey the law of the excluded middle: the statement that the cat is on the mat must be either true or false under any conditions.

9. ON THE ACCEPTABILITY OF SCIENTIFIC THEORIES

Considerations to this point suggest that we would face insurmountable difficulties should we attempt to attribute probability values – whether or not under the name of 'degrees of confirmation' – to the various expressions of modern science. In the case of scientific theories, each of which postulates the existence of a hitherto unknown realm of entities whose behaviour is to be responsible for certain empirical regularities, we must first determine what is to *count* as evidence for or against the theory in question.

As has been pointed out by Harré,[11] this involves a consideration of more than simply how the theory should manifest itself empirically, for one also must take into account or treat as evidence the reasonableness of assuming the mechanism the theory posits as actually existing and being responsible for the regularity. As mentioned in Chapter 4, Bohr's theory of the atom, for example, can well account for the spectral lines of hydrogen gas; but in so doing it suggests that electrons within the atom move from one orbit to another without occupying the intervening space. This contravenes a particular form of the principle of the continuity of space, and its doing so cannot be ignored when considering the likelihood of what the theory suggests actually being the case.

But even if we recognise this fact and say, for example, that one of two empirically equivalent theories is more likely to be true if it does not contravene any principles, we still do not know how to weight the two kinds of factor – i.e. empirical evidence and accordance with principles – particularly with respect to each other. Their difference seems to be fundamentally of a qualitative sort, and it is virtually

---

[11] (1970b), p. 166.

impossible to imagine how one might arrive at a weighting procedure which is not highly arbitrary.

Also, as emphasised in Chapter 5, due to the idealisational nature of scientific theories, it is not at all clear what would have to be the case for us to say of them that they are *true*. The kinetic theory of gases treats molecules as though they were perfectly spherical. Is anything in reality perfectly spherical? Taking the idealisational nature of theories into account we would simply have to say, granting that they have truth-values, that all theories are false; and as a consequence it would be pointless to attempt to apply probability values to them.

What is done instead, as is reasonable, is that one speaks of the *acceptability* of theories, either alone or as compared to other theories. And it is not imagined that this acceptability comes in degrees computable with the aid of an algorithm, but that it is qualitative and determined by the informed judgement of scientists.

10. ON THE CONFIRMATION OF EXPERIMENTAL LAWS

It may be thought, however, that while we may not be able to speak of the quantitative probability (or degree of confirmation) of scientific theories, we still might be able to speak of the quantitative probability of (the truth of the expressions of) empirical laws. But here again we face the problem of determining what is to count as the entity in question being true. No empirical laws are strictly true, or hold exactly, but all are relatively applicable within a certain range of experimental conditions, and less applicable outside that range. Thus if we admit that such laws are not entirely false, the requirement stressed in Section 8, that there must be no vagueness concerning the statement with regard to which the probability assignment is made, is not met in this case.

Another aspect of empirical or experimental laws that must be mentioned here is that they are determined in accordance with scientific principles, the most notable being the principle of the uniformity of nature. It is for this reason that an experiment need be repeated only a few times, mainly to ensure that no mistake has been made in the way it has been conducted, in order for its results to be established and incorporated into the body of scientific fact.[12] Thus, though the (qualitative) probability of the results of the experiment being correct increases when the results are repeatedly obtained by re-staging the experiment, after only a few such re-stagings these results are accepted as certain, and no longer as merely probable. And it is in this sense that one speaks of empirical laws as being confirmed.

11. ON THE APPLICABILITY OF SCIENTIFIC PRINCIPLES

Scientific principles, whether tacit or explicit, set guidelines for the conducting of scientific enquiry, both empirical and theoretical, or on both the nominal and real levels. That they should be considered either true or false is also problematic. Though it is to be admitted that they may face counter-instances, in such cases we are more inclined to question whether they have been properly refined, or are being applied to a suitable subject-matter. Thus we might say, for example, that the fundamental principles of modern science are better applicable to physical entities and phenomena than to biological or social-scientific ones, rather than that they are true or false.

Another question transcending the true/false dichotomy is that of *relevance*. A particular set of principles may apply very well to some domain ('may be close to being true'), but there may nevertheless be no interest in applying them. Or, on the other hand, they may not apply at all well, while the attempt to apply them proceeds with vigour.

But even if we were to speak in terms of principles' being true or false, it is impossible to conceive of real states of affairs which actually show them to be false, for we can never be sure that we know everything of relevance to the situation in question that would warrant our drawing such a conclusion. Thus scientific principles are seldom discarded entirely, but are more often set aside in those situations where they apparently do not apply.

As a consequence of this, if we consider applying the notion of quantitative probability or confirmation to scientific principles, we

---

[12] In this regard, cf. above, p. 78, n. 4 and accompanying text.

face difficulties at every turn. We cannot say that what appears to constitute evidence for or against a particular principle necessarily does so, for our lack of total knowledge of the relevant situation may mislead us on that score. And so we cannot give a numerical weight to the evidence, nor can we indicate the methodological grounds which tell us how to assign a probability value to the principle itself.

Consideration of the actual natures of scientific laws, theories and principles suggests not only that a project aiming to determine their quantitative probability or 'degree of confirmation' could never be realised, but that the very idea of providing an algorithm for making such determinations is based on an overly simple conception of the nature of science.

In the next, penultimate, chapter, we shall return to the issue of empiricism vs. realism, and consider it against the background of the main content of this work.

CHAPTER 9

EMPIRICISM VS. REALISM REVISITED

The debate in the philosophy of science between empiricism (positivism) and realism is not so much a debate concerning how science is or has been practised, as one concerning how it *ought* to be practised. Empiricists, for their part, view the aim of science as the affording of truth, and want therefore to exclude from science any activity of a hypothetical nature. Realists, on the other hand, see the aim of science as concerning understanding as well as truth, and view informed speculation about the nature of the real world as a worthwhile attempt to obtain such understanding.

1. THE HISTORICAL DEBATE

As regards the historical debate treated in Chapter 1, as to whether (modern) science ought to investigate the causes of phenomena as well as the phenomena themselves, the present study supports the view of William Whewell and the other realist contributors. From an epistemological point of view, to limit science to the investigation of phenomena and their formal relations would be to deny the relevance of the principles of substance and causality to modern science – which have here been shown to be part of its essence. This would mean giving up the quest of making the phenomena understandable in a broader perspective, and at the same time losing a source of inspiration regarding how new phenomena or laws might be discovered. Furthermore, important scientific advance has in fact depended on the realist approach, such as in the case of the establishment of the atomic theory of matter and the genetic theory of heredity, so that a characterisation of modern science as it actually has been practised which did not take account of its realist aspect would be incomplete to say the least. Here it is to be noted that the realist alternative in fact encompasses its opponent, and does not exclude it, while the opposite is not the case. On the view of modern science presented in this study, in which the enterprise emanates from the uniformity, substance and causality principles, the investigation of the causes of phenomena in order to understand the phenomena themselves is as important an aspect of science as is their discovery.

What may account for the persistence of the empiricist approach is philosophers' general neglect of the importance of obtaining understanding of phenomena over and above the attaining of knowledge regarding them.[1] A perusal of the history of Western philosophy would reveal that many influential philosophers, rationalists as well as empiricists, have seen the obtaining of *certainty* as the primary goal of epistemological endeavours.[2] Even Kant was concerned to have certain knowledge on the transcendental level.

It may be argued however that Kant's view, for example, evinces a lack of sophistication in its understanding of the relation between the transcendental and the factual. These conceptual levels play different roles in epistemology, and to say that both constitute forms of knowledge misses the more essential difference in their functions. Both in philosophy of science and in a broader philosophical context, the ultimate defence of the transcendental is not that it is indubitably certain, but that without it nothing makes sense.

Facts are only facts given certain presuppositions regarding what it means to be a fact. These presuppositions are, in the present context, the core principles of modern science, especially that of uniformity. By conceptual necessity such principles must lie at a more fundamental level than do the facts which presuppose them.

For modern science, the transcendental need not and in fact does not consist of apodeictic truths, but of a conceptual outline of an ideal reality, the details of which are to be filled in by investigating the reality in which we live as though its fundamental nature were that of the ideal. Thus the principles of modern science are transcendental *relative* to the enterprise, and are susceptible of modification from a position outside it.

---

[1] Cf. in this regard N. Maxwell's suggestion, that "[t]his standard empiricist view seriously *misrepresents* the true intellectual aim of science. The aim of science is not merely to discover truth *per se*, nothing being presupposed about the nature of the truth to be discovered. A basic aim of science is to improve our *understanding* of the world." (1984, p. 96).

[2] Cf. above, p. 13, n. 8 and the accompanying text.

The above of course implies that science necessarily involves metaphysics – a situation positivists are unanimous in their desire to avoid. But the transcendental is not the only metaphysical aspect of science, for, as we have seen, the transcen*dent*[3] is also present, in the form of speculative theorising. In delving into the unknown in an attempt to explain the known, the theorist employs transcendent thinking – a rather obvious point which suggests a surprising naivety on the part of the logical empiricists in wanting to retain the theoretical aspect of science while ridding science of metaphysics.

Another aspect of the empiricism-realism debate that deserves further attention is the empiricist inclination, most clearly expressed in Mill, to identify causality with uniformity. This identification has seldom been explicit, and has been all the more pernicious when implicit. Throughout the history of modern philosophy leading philosophers, including Hume and Kant, have at least on occasion seen the notion of causality as implying that of determinism, a predilection which is still widespread today, as is manifest for example in debates concerning the foundations of quantum mechanics. Until the notions of causality and uniformity are clearly separated, however, it is impossible to carry on a meaningful discussion of the appropriateness of empiricism or realism with regard to the conducting of the enterprise of science.

But this is only one point, albeit a vital one, where the current discussion fails to be enlightening due to its ignorance of fundamentals. As pointed out in Chapter 1, even more important in this regard is the unwitting acceptance by most influential contributors to the debate of the framework of the logical empiricist conception of science,[4] itself arrived at with no consideration being given to the true

---

[3] For Kant's conception of the distinction, see his (1783), Ak. 373n.: "[T]he word 'transcendental'... does not signify something passing beyond all experience but something that indeed precedes it *a priori*, but that is intended simply to make cognition of experience possible. If these concepts overstep experience, their use is termed 'transcendent,' which must be distinguished from the immanent use, i.e., use restricted to experience."

[4] That this framework has not been repudiated by modern analytic philosophy is noted by Blackmore: "[F]or people who favour an *indirect* philosophy [such as realism] the similarities between logical positivism and analytic philosophy continue to appear much more numerous and important than the differences, and going back further in time, the resemblances with Comte's positivism and the philosophies of

nature or even the existence of the realist alternative. And, from the point of view of the present study, one of the basic blunders of logical empiricism, which is still with us today, is the unquestioned assumption that modern science has a structure determined by formal logic. Not only has this view received no supporting argument from those who accept it, but, as has been argued by Whewell, Campbell and Harré, it is quite simply mistaken.

As mentioned in Chapter 1, it is natural on an empiricist conception to see the relations of epistemological relevance to science as being formal relations between phenomena, since causal relations are excluded on an empiricist view. From this it is a comfortable step to take the structure of science itself as being a formal (or formalisable) one, either that of logic or of mathematics. Such a move is also supported of course by the certainty formalism purportedly affords. But just as conceiving of the subject-matter of science as consisting of phenomena and formal relations between them misses the whole of the substantial and causal aspects of science, so conceiving of science as having a purely formal structure misses the fact that it is an enterprise involving contact between ideas and the real world through operations of measurement, and is aimed at providing a type of knowledge and understanding whose nature is determined by the particular core principles it adopts. As demonstrated in Chapter 2, due to its core principles scientific thinking is on the one hand not so rigid as logical thinking, and on the other is much more specific as regards the nature of its object.

2. THE SUPERVENTION OF EXPERIENCE

The difference between the empiricist and realist conceptions can be characterised as fundamentally a difference of *attitude*. Where the empiricist is concerned with reality as it is experienced, the realist is concerned with reality as it is independently of how it is experienced.

It has generally been assumed that at least the empirical aspect of modern science is in keeping with the empiricist conception; but as

---

Berkeley and Hume continue to seem much more widespread and important than the differences." (1983, p. 33).

has been argued in Chapter 3 this is actually not the case. The fact that empirical scientists employ measurement in their acquisition of knowledge demonstrates that their interest is in reality as it is in itself, not reality as they experience it, and that as a consequence they themselves are realists in their orientation at least to this extent. This being the case is somewhat ironic when one considers the influential campaign carried on during the past century in the name of positivism (behaviourism), having as its aim the employment solely of quantitative empirical methods in the social sciences. If the hope has been to eliminate metaphysics in this way, it is a vain hope, since measurement presupposes a transcendental realm. What might possibly be eliminated by such a procedure however is the transcendent realm of theorising which, in the social sciences, involves speculation regarding the motives and intentions of human beings. But the problem from a broader epistemological perspective is that what is lost in such a move is an *understanding* of how social change comes about as a result of causes emanating from the minds of social actors.

If we look at the development of modern science during the last century we see a widespread move towards this form of positivism – i.e. towards a disinclination to posit theories regarding the causes of phenomena – in both the social and the natural sciences. As regards the latter, this orientation has been extremely influential in the form of the Copenhagen interpretation of quantum mechanics, a result of the apparent impossibility of framing a conception of quantum events which in the first place is in keeping with a strict form of the uniformity principle (due to indeterminacy) and with the proximity principle (due to non-locality), and in the second accords with our prescientific notions of what is physically possible. A common underlying motive may exist for all cases, namely that of avoiding being mistaken by restricting one's attention to areas in which one feels assured that one can obtain objective knowledge. In the social sciences this may be conjoined with the thought that *real* (physicalist) scientific methodology does not involve reference to such ethereal entities as human minds, and the belief that 'real science' does not involve understanding. But as has been argued by the realist commentators cited in Chapter 1, such an orientation not only leaves questions as to the how and why of phenomena unanswered, but is stultifying even as regards the acquisition of new knowledge in those areas where it is believed that it is obtainable.

3. ONTOLOGY VS. EPISTEMOLOGY

The strong influence of positivist thinking on the actual conducting of science during the past century has also had an effect on the question as to the extent to which the principles of science have in fact been qualified during this period. One of the theses of the present work is that such changes in the scientific enterprise as have been brought about by, for example, relativity theory and quantum mechanics are the result of changes in the principles of science. But a closer look at this phenomenon indicates that the change is more radical when seen from an empiricist than from a realist point of view. One reason for this is that much of the fundamental revision that has actually occurred has been the result of insights regarding particular epistemological limitations, limitations which for the empiricist but not for the realist are taken as having direct ontological implications. (Another reason is the empiricist conflation of uniformity and causality, the epistemological inability to discover a relation of strict uniformity being taken as proof of the failure of the ontological principle of causality.)

Here we might consider some examples of empiricist over-reaction against the principles of science, beginning with their interpretation of the special theory of relativity. Against the background of a conception of time-determination based on the passing of signals between spatially separated bodies, special relativity suggests that the simultaneity of events on bodies moving relatively to one another is itself relative. Commentators of an empiricist or positivist orientation take this immediately to constitute a (true) claim regarding the world – that in reality events on relatively moving bodies cannot be simultaneous[5] – thereby constituting an important qualification of the ontological principle of the uniformity of nature. From

---

[5] Even Čapek, who cannot be considered an empiricist, makes this error where he says: "We should avoid interpreting the term 'moment' in its cosmic prerelativistic sense; for we know that there are no world-wide instants." (1961, p. 344).

a realist point of view, on the other hand, the relativity of simultaneity suggested by the special theory is an epistemological limitation, not an ontological one, and thus does not imply a qualification of the ontological uniformity principle, but a limitation of scientific epistemology.

Similarly, the orthodox positivistic interpretation of quantum mechanics assumes the epistemological limitations regarding the simultaneous determination of the position and momentum of a body to have the ontological implication that a body cannot in reality have both position and momentum simultaneously – a state of affairs which would imply an inherent vagueness in the principle of the uniformity of nature. This assumption was of course firmly rejected by the realist Einstein, who argued that the inability to determine position and momentum simultaneously, rather than have ontological implications, reflected a weakness in quantum mechanics.[6] In the same vein we may refer back to our argument in Chapter 8 regarding the propensity interpretation of probability. The inability epistemologically to predict the path followed by a particular electron other than probabilistically, even if this inability is one of principle, does not imply that the path is not ontologically predetermined.

A final example to be mentioned, only because it has been often referred to recently as constituting a counterexample to the uniformity principle (or to the causality principle conceived as necessarily implying uniformity), is that of chaos theory. As regards this 'theory' it may first be noted that, in being basically mathematical, it has no natural place in modern science. (It is similar to, e.g., game theory in this respect.) In fact, due to this characteristic, chaos theory, unlike theories of modern science, lacks an ontology. It consists rather of conceptualisations involving formal reiterations which, represented graphically, bear certain resemblances to particular em-

---

[6] Cf. Einstein (1949a), p. 672: "[I]f the statistical quantum theory does not pretend to describe the individual system (and its development in time) completely, it appears unavoidable to look elsewhere for a complete description of the individual system. ... With this one would admit that, in principle, this scheme could not serve as the basis of theoretical physics. Assuming the success of efforts to accomplish a complete physical description, the statistical quantum theory would, within the framework of future physics, take an approximately analogous position to statistical mechanics within the framework of classical mechanics."

pirical phenomena. Thus one of its basic presuppositions in its application to reality is that certain empirical systems operate in the same way as the formal systems. So just as one cannot predict the development of the formal system from mere knowledge of its graphical representation at one or more points in time, it is suggested that one cannot predict the development of the corresponding empirical system (and thus that its development is 'chaotic'), since it is assumed to develop in the same way as the formal system. But here we note two points. First, whether one is able to predict the development of a system is an epistemological question distinct from the ontological question of whether the system actually develops deterministically; and second, in the present case not only is the development of the relevant system conceived to be strictly determined by particular non-linear equations, but given knowledge of these equations one should in principle also be able to predict the way in which the system will develop. Here again then we have an instance where the limitation is an epistemological and not an ontological one, furthermore in this case a limitation which applies only to the empirical level.

Thus we see that some of the more revolutionary changes of twentieth-century science involve an alteration of epistemological principles rather than of ontological principles, and further that the suggestion that such changes themselves represent ontological insights is misguided, as has been argued by Einstein. What this means regarding e.g. quantum mechanics is that the 'theory' itself is inadequate in a certain respect; and that, due to the fundamental nature of the theory, the inability of physicists to overcome this inadequacy during the more than seventy years since the theory's inception constitutes an impasse for the development of the central discipline of modern science. More generally, these various restrictions of the epistemological principles of modern science, rather than indicating its flexibility and lack of reliance on ontological principles (as the empiricists would have it), indicate instead a weakness in the epistemological activity of modern science itself.

4. UNDERSTANDING VS. KNOWLEDGE

While realists accept both the existence of a world independent of human experience, as well as the meaningfulness of thinking about that world, they need not believe one can have knowledge of such a world. What is being urged in the present work is that realism with respect to science be conceived of as embodying two kinds of investigative procedure. The one is the acquiring of *knowledge* of empirical facts by measurement; and the other is the *making sense* of those facts by conceiving of a particular ontology which might ground them.[7] This conception of realism would apparently differ from the conception of some realists, on whose view the obtaining of knowledge of the entities referred to in the ontology is or ought to be the primary aim of science.

While the above view is not excluded by the present work, there are a number of reasons why it has not been adopted here, all of which coalesce in a particular conception of the epistemological enterprise called modern science, in which theory construction is paradigmatically hypothetical. One reason is that claiming that scientists can obtain knowledge of entities which are at one time conceived of as transcendent, such as atoms, says more than is necessary to vindicate realism in science. Another is that the view that emphasises the obtaining of knowledge of ontological entities tends to collapse the theoretical into the empirical, thereby obfuscating the distinctions between the hypothetical and the factual, and between explanation and description. And a third is that, as has been appreciated by Whewell,[8] the empirical has a particular priority over the real: the empirical constitutes the point of departure – that with which one is originally confronted – while the real is posited only afterwards, first in order to explain the empirical.

In the context of thinking about modern science, realism obtains a more distinct position in the present work than it would if it were seen as another variation on the theme of striving for the truth. In a broader philosophical context, it may be said that the knowl-

---

[7] Cf. ibid., p. 669: "The justification of the constructs, which represent 'reality' for us, lies alone in their quality of making intelligible what is sensorily given."

[8] See above, p. 96, n. 1 and accompanying text.

edge/understanding distinction emphasised here is a particular manifestation of the factual/transcendent distinction, a distinction which,
even if considered to be relative rather than absolute, cannot be lost
sight of in epistemology.

In the next and final chapter we shall look at modern science in an
historical perspective, and consider what its future might be.

CHAPTER 10

MODERN SCIENCE AND THE FUTURE

In the debate between realism and empiricism the question of the viability of modern science is seldom raised, the issue being rather one of how modern science ought best be conducted. In fact few of the contributions to the debate, whether realist or empiricist, even recognise that fundamentally different alternative forms of science are possible, the belief being rather that with the Scientific Revolution of the seventeenth century, if not earlier, humankind hit upon the one true *scientific* path to knowledge, the task now being that of continuing to follow that path in the best possible way.

1. A PARTICULAR ENTERPRISE EMANATING FROM PARTICULAR PRINCIPLES

Concomitant with the view that (modern) science constitutes the only truly viable epistemology is the view, expressed e.g. by Quine,[1] that science as we know it is actually nothing other than a continuation of common sense. This has the implication, among other things, not only that there is something inherently right about science, but also that its adoption on the part of humankind was in some sense inevitable.

While it may be that there was some kind of inevitability in our intellectual development passing on to modern science, it need not be necessary that we remain at that stage. As the attempt has been made to show in this work, modern science is not simply the elaboration and refinement of common sense, but constitutes a very particular enterprise whose limits are set by the principles from which it emanates. Neither these principles nor the physicalistic interpretation they have been given by modern science are inviolable, however, and to a large extent both have been adopted as a result of the par-

---

[1] Quine (1951), p. 45; regarding this question, see further Dilworth (2004).

ticular interests of the research community. If those interests were different, what we consider to be science could well be centred either on other principles or on the same principles but under another interpretation. To obtain a wider view on modern science in which such a possibility might be realised, it may be worthwhile to consider the enterprise in a perspective covering the whole of the history of humankind.

2. THE REVOLUTION FROM MYTHOPOEIC THOUGHT

To the best of our knowledge our first modern human forefathers (*Homo sapiens sapiens*), the Cro-Magnons, appeared on the earth as recently as about 200,000 years ago. They were a nomadic people who had the use of fire and eventually the bow and arrow, and practised various magical or religious rites. At the end of the last glacial period, about ten thousand years ago, we humans first left the nomadic way of life and became engaged in agriculture and the rearing of domestic animals. This led to the creation of cities and civilisation as we know it, first in Egypt and Mesopotamia.

Common to nomadic life and early civilisation was the mythological nature of people's worldviews. Virtually all humans living three thousand years ago believed that the major events in the world were caused by gods,[2] one or more of whom perhaps created the world itself. As has been described by some archaeologists, people's relation to the world in which they lived was an I-Thou relation,[3] and the

---

[2] Cf. Hübner (1988), p. 108: "Myth is founded on numinous experience. This means that it interprets the objects of experience as numina, i.e. as signs of a divine effect or presence. The same is true for the interrelation of objects, whether they be rules or whether they arise spontaneously. Rules, such as e.g. rhythms or periodical appearances of nature, are from the mythical point of view only repetitions of divine archetypes, whereas all spontaneous events are reduced to the fateful influence of a god."

[3] Cf. Frankfort et al. (1946), p. 12: "The ancients, like the modern savages, saw man always as part of society, and society as imbedded in nature and dependent upon cosmic forces. For them nature and man did not stand in opposition and did not, therefore, have to be apprehended by different modes of cognition. [N]atural phenomena were regularly conceived in terms of human experience and ... human experience was conceived in terms of cosmic events. ... The fundamental difference

belief was that through praying or sacrifice one could influence the gods and thereby bring about changes on earth.

Some 2,600 years ago was initiated what may be considered the greatest intellectual revolution in the history of humankind, when it was suggested for the first time that major events in the world, such as earthquakes, were not caused by irate gods, but emanated from the inanimate world itself.[4] The person with whom we can associate the first step in this revolution is Thales, who, among other things, attempted to explain earthquakes by likening the earth to a ship floating on water, which quakes when the movement of the water causes it to rock. With Thales we have both the birth of Western philosophy as well as the first known instance of a causal theory of the modern-scientific form.

But the intellectual revolution instigated by Thales was not one leading directly to modern science, but rather to what we might term philosophical thought. What distinguishes philosophical thought generally from mythopoeic thought is not its being inherently more rational than the latter,[5] but its positing of 'natural' rather than su-

---

between the attitudes of modern and ancient man as regards the surrounding world is this: for modern scientific man the phenomenal world is primarily an 'It'; for ancient – and also for primitive – man it is a 'Thou.'"

[4] "Primitive thought naturally recognised the relationship of cause and effect, but it cannot recognise our view of an impersonal, mechanical, and lawlike functioning of causality. For we have moved far from the world of immediate experience in our search for true causes, that is causes which will always produce the same effect under the same conditions. [T]he primitive mind...looks, not for the 'how,' but for the 'who,' when it looks for a cause. [T]he gods as personifications of power among other things fulfil early man's need for causes to explain the phenomenal world." ibid., pp. 24, 26.

[5] Cf. Hübner (1988), pp. 109–110: "[M]yth possesses rationality. Nevertheless myth and logos are thoroughly different. The essence of logos is the *lógon didónai*. This means the challenge to reduce the variety of the given to ultimate principles. So e.g. the pre-Socratic philosophers, who initiated logos, derived the totality of appearances from water, fire, air etc. ... By progressively reducing the given to uniform, purely conceptually formulated foundations, the different scopes of numinous beings were being deprived of their ground. The way was opened for the destruction of myth."

pernatural causes of phenomena,[6] and at the same time being speculative rather than dogmatic in its views.[7]

3. THREE STREAMS IN GREEK THOUGHT

While Thales suggested a way of conceiving of reality in which the
causes of change were other than supernatural, his conception was
not the only one, but was followed by others, one of which took
causes as being fundamentally formal, and another of which took
them to be goal-directed but without the support of a human-like
supernatural agency. In keeping with the views of Burnet, Northrop
and von Wright,[8] we here suggest that there were basically three
streams in Greek philosophical-scientific thought, one physicalist in
its orientation, one formal, and one biological.

The *physicalist* stream began with Thales and was developed
mainly by the Presocratic philosophers. Its basic presuppositions
were essentially those of modern science. Already in Thales, for example, there was an expression of the principle of perpetual physical
substance, the substance being water, and everything else being a

---

[6] "It remained for the Greeks, with their peculiar *intellectual* courage, to discover a
form of speculative thought in which myth was entirely overcome." Frankfort et al.
(1946), p. 248. In this regard see also Burnet (1914), p. 29: "[I]t was just this non-
religious use of the word 'god' which made it possible for the Milesians to apply it
to their primary substance and their 'innumerable worlds.' That way of speaking does
not bear witness to any theological origin of Greek science, but rather to its complete
independence of religious tradition. No one who has once realised the utter secular
character of Ionian civilisation will ever be tempted to look for the origins of Greek
philosophy in primitive cosmogonies." See also e.g. Burnet (1920), pp. 14–15; Bailey
(1928), pp. 24 and 47; Northrop (1931), p. 3; Frankfort et al. (1946), pp. 253–256, 260–
262; and Kahn (1960), pp. 116–117: "In the historical experience of Greece, Nature
became permeable to the human intelligence only when the inscrutable personalities
of mythic religion were replaced by well-defined and regular powers. . . . The strife of
elemental forces is henceforth no unpredictable quarrel between capricious agents, but
an orderly scheme in which defeat must follow aggression as inevitably as the night the
day." Of course the atomist view in particular is explicitly atheistic: see e.g. Lucretius,
pp. 31, 92 and 177.
[7] Our last point is expressed by Frankfort et al. thus: "[M]yth claims recognition by
the faithful, not justification before the critical." (1946, p. 251).
[8] See Burnet (1914), pp. 11, 44; Northrop (1931), Ch. 1; von Wright (1986), p. 58.

form of water.[9] The physicalist line took a particularly important step with Parmenides who, concentrating on the idea that one cannot conceive of what is not, argued that what exists is one and change is impossible. The problem of explaining change thereby became a focal point of philosophical thought, a problem rather soon to be 'solved' by Leucippus.

Leucippus' solution to the problem posed by Parmenides was to admit the non-changeability of that which exists, but to add that not-being (space) also exists, and that being was not one but many, consisting of an infinite number of invisible atoms. Thus, on Leucippus' view, all change consists in the motion of perpetually-existing physical atoms in infinite space. This motion is also perpetual, and all change is deterministic, consisting either in the rectilinear motion of the atoms, or in their acceleration through collision. Thus in ancient atomism we have the three fundamental principles of modern science in their pristine form: the purely deterministic principle of the uniformity of nature, the principle of substance as consisting of perpetually existing physical atoms, and the contiguity principle of causality according to which all change of state is the result of physical impact.

The *formal* stream in Greek thought began with Pythagoras and was carried further, via Socrates, by Plato.[10] Like the physicalist stream, the formalist line adopted the perpetuity principle of substance, the substance being formal rather than physical, however. With Pythagoras, the substance consisted of number; and with Plato it became the ideal Forms in comparison with which the world of experience is but a shadow. These Forms, like the physicalists' atoms, constituted true reality, inaccessible to the senses.

The third stream in Greek thought is the *biological*, which may be considered to have begun with Hippocrates, who emphasised the notion of the whole, and to have been fully developed as a philoso-

---

[9] Here we note how, as early as the time of Thales, ontological principles were not empirical generalisations, as has been noted by Northrop (1931, pp. 4–5), but synthetic a priori presuppositions.

[10] In this regard, see Burnet (1914), pp. 89, 91–92, 152–155, 314. Cf. also Jammer (1954, p. 12): "With Plato physics becomes geometry, just as with the Pythagoreans it became arithmetic."

phical system independently by Aristotle (though there were other aspects to Aristotle's philosophy besides this – most notably his logic). On this line reality and change are conceived in analogy to the biological, either on the level of organisms or on the level of species or higher taxa. The explanation of things' being as they are consists in indicating that, due to their essences, they occupy a particular position in a taxonomy, while the explanation of their changing consists in indicating that, due to these same essences, they have a propensity to develop in a particular way. Here the biologically based notions of generation and organisation are central. The paradigm of change is that of the birth and growth of an organism from a seed, its mature state being that which the seed itself strives to attain. The cause of change on this philosophy is thereby fundamentally teleological[11] – though both the form *and* the matter of the entity in question are necessary aspects of the cause – with the present state of affairs constituting what is potential, and the final state (*entelecheia*) what is actual. Thus, unlike for the atomists, *all* motion (change) is caused, and furthermore its cause is internal to that which moves. Earth and water, for example, being heavy objects, naturally seek the centre of the world as their resting place, while fire and air, being light, naturally seek the heavens. Both prime matter (the substratum) and the formal may be seen as unchanging, though the formal, unlike for Plato, does not occupy a transcendent world of Forms, but (apart from in the heavens) exists in empirical reality in the same way as do unchanging species.

4. CHRISTIANITY, PLATONISM, ARISTOTELIANISM AND THE SCIENTIFIC REVOLUTION

Though the Christian worldview differs from what are normally considered mythological worldviews in that the latter involve belief in a variety of gods while there is only one Christian god (the Trinity united in God), Christianity is nevertheless an essentially dogmatic system of beliefs according to which individuals can influence worldly events through supplication to a supernatural agency. While

---

[11] As is suggested by Guthrie (1965, p. 416).

during Greek times philosophers were in a small minority in their various conceptions of reality as not being directly influenced by the supernatural, with the appearance of the Christian era virtually everyone in Europe, philosophers included, subordinated non-spiritualistic conceptions to spiritualistic ones. Thus, to the extent that philosophical thought existed in Europe from the beginning of the Christian era up until the seventeenth century and even later, it was largely moulded to fit the Christian faith.

Furthermore, when philosophy did manifest itself during this period, it consisted essentially in a return to, or continuation of, Greek ideas. In fact, as will be argued here, the three streams of Greek philosophical thought have shaped virtually the whole of Western intellectual development – philosophical and modern-scientific – right up to today.

The first stream to make itself felt after the political decline of Greece was the formal line in the form of Platonism and Neoplatonism in late antiquity and the Middle Ages. One can say that some form of Platonic thought dominated philosophy, when it was practised, from Roman times until the thirteenth century.[12] Among the most notable adaptations of Platonic thinking to Christian dogma is that of St. Augustine (354–430). Of the three lines of Greek thought, the formal is perhaps the most amenable to such an adaptation. The perfect world of Ideas, with Plato's notion of the Good as the highest Idea, both is similar to and has in fact influenced the Christian notions of heaven and God. What the Platonic line lacked however, from the point of view of its potentially constituting a fruitful science, was a clear link to empirical reality, as well as an integral notion of change. Its focus was too firmly fixed on a transcendent world of static Forms.

The next stream to have an influence generally in Europe was the biological, through Aristotle's philosophy. Though Aristotle's works, other than those in logic, had largely disappeared from

---

[12] Cf. Jammer again: "This identification of space and matter, or, in the words of later pseudo-Platonic teachings, of tridimensionality and matter, had a great influence on physical thought during the Middle Ages. For although Aristotle's *Organon* was the standard text in logic, Plato's *Timaeus* was succeeded by Aristotle's *Physics* only in the middle of the twelfth century." (1954, p. 14).

Europe during the early Middle Ages, they had been preserved in the Middle East, becoming available in Europe from about the mid-twelfth century. They strongly influenced the writings of Albert the Great (c. 1200–1280), and were synthesised with Christian Neoplatonism in the work of his pupil, Thomas Aquinas (c. 1224–1274). In keeping with the link to empirical reality implied by Aristotle's biologically inspired view of reality, Aquinas' philosophy emphasised the acquisition of knowledge via sense experience, and was greatly indebted to Aristotle's teleological view in its theory of physical reality. Furthermore, Aquinas' thought was also teleological in his practical philosophy.

In the sixteenth and seventeenth centuries the Aristotelian philosophy came under serious attack. First was the presentation and ultimate victory of the Copernican (Aristarchean) conception of the solar system, which removed the earth from the centre of the universe; and second was the Galilean-Newtonian physics, which refuted the idea that all motion is caused, and treated the earth and the heavens as being the same kind of existents obeying the same laws. Both of these triumphs for modern science were triumphs for the physicalist stream of Greek thought – the latter even moreso than the former.[13] More particularly they were triumphs for the worldview of atomism, a worldview which had in fact been adopted by virtually all of the leading figures of the Scientific Revolution. The viewing of the sun rather than the earth as being in the centre of the universe constituted a first step towards seeing the sun as one star of infinitely many in an infinite space without centre, as is in keeping with atom-

---

[13] "In fact, both Copernicus and Kepler were ardent adherents of the mathematical theory of nature. It was the ideas and discoveries of Galilei, the physicist, which produced the real revolution – the philosophical revolution, for they necessitated the rejection of the traditional dominant theories of first principles. They were incompatible, not merely with Platonic and Aristotelian science, but also with the Platonic and Aristotelian, or mathematical and functional, philosophies of science. It is for this reason that Galilei ushered in a new day, rather than a mere radical modification in the secondary principles of an old one, when he made forces and the motions of masses the fundamental concepts of science. ... Thus, with the discoveries of Galilei and the generalisation of those discoveries by Newton and Laplace, the physical theory of nature became the dominant philosophy of the Western World, for the first time in its history." Northrop (1931), pp. 32, 33.

ism.[14] And the affirmation of the principle of inertia, which states that change of state rather than motion *per se* is caused, such a cause being an external (efficient) one, reinforced the atomist view that motion is perpetual.[15]

The metaphysics underlying the Scientific Revolution was that of early Greek atomism. As an historical thesis, this view is supported by the whole text of the present work, the three principles in terms of which modern science has here been explained, together with their physicalist interpretation and the contiguity principle of efficient causality, being presupposed by early atomism.[16] It is with *atomism* that one obtains the notion of a *physical* reality underlying the phenomena, a reality in which *uniform causal* relations obtain. This metaphysics was at the time of the Scientific Revolution first laid out by such philosophers as Bruno (1548–1600), Campanella (1568–

---

[14] Cf. Čapek (1961, p. 19): "Thus when Giordano Bruno attacked the finite universe of Aristotle, he was at the same time destroying Aristotelian physics; in returning to the infinite space of the ancient atomists, he was reasserting the homogeneity of space, which was incompatible with the doctrine of four elements and their 'natural places.'"

[15] Cf. ibid., pp. 72–73: "[F]orce is needed to keep a body going in the Aristotelian universe, while this is not so in the world of Galileo and Newton. We have sufficiently emphasised how inferior Aristotle was in this respect to the early Greek atomists, who lacked only more sharply defined kinematic notions to arrive at a correct formulation of the law of inertia and that of quantity of motion."

[16] For independent support, see e.g. Northrop: "[S]cience took over the physical theory of nature as Leucippos...stated it." (1931, p. 10); Dijksterhuis: "Ancient atomism...had essentially been revived in the corpuscular theory. ...[T]he conception of the world which seemed to result almost inevitably from the mechanistic philosophy was in effect no different in the seventeenth century from what it had been in Antiquity. [T]he materialist metaphysics was gaining ground." (1959, p. 453). Cf. also Čapek: "The only difference between Greek atomism and nineteenth-century physics was that the latter had incomparably more efficient technical and conceptual tools at its disposal than Democritus and Leucippus; the vague necessity (*ananke*) of Greek atomism has been replaced by the precise conservation laws of modern dynamics. Fundamentally, however, the basic conceptions were the same. This was the deep historical reason why the birth of modern science occurred simultaneously with the revival of atomism by Bruno, Bacon, Gassendi, and others." (1961, p. 123); and Hutten: "The Greek atom is, of course, not the same thing as the modern atom, but it was only when this concept was rediscovered by Galileo as the mechanical 'particle' that science began in the sense in which we understand it today." (1962, p. 75).

1639) and Gassendi (1592–1655), at the same time and later being adopted by such experimentalists as Galileo, Boyle, Huygens and Newton. That the world consisted essentially of matter causally interacting with other matter by contact became the intellectual conviction of the times, adopted even by medical researchers such as Harvey (1578–1657) and social theorists such as Hobbes (1588–1679).[17]

But while we can say that atomism constituted the metaphysics of the Scientific Revolution, the empirical aspect of modern science, consisting in the determination of natural laws via measurement, had already been manifest independently of atomism in the work of such ancients as Pythagoras, Archimedes and Heron. One might here ask why the Scientific Revolution did not occur then with Pythagoras or Archimedes. Part of the answer lies in their not having made the step to being able to measure change or motion. But more important is the fact that their methodology was not grounded in a metaphysics which both took motion as basic, and at the same time could make sense of the results they obtained in terms of a particular ontology. What made the Scientific Revolution truly distinct, and Galileo rather than Archimedes its father, was that for the first time this empirical methodology was given an ontological underpinning; or, better, that for the first time it was realised that the transcendental principles of the metaphysics of atomism could, via this methodology, make clear contact with empirical reality.

Thus, where the properties of atoms were quantities and not qualities, as was necessitated by their all consisting of qualitatively the same physical stuff,[18] Galileo and Newton, starting from this atom-

---

[17] As expressed by Burtt: "From now on it is a settled assumption for modern thought in practically every field, that to explain anything is to reduce it to its elementary parts, whose relations, where temporal in character, are conceived in terms of efficient causality solely." (1924, p. 134).

[18] As expressed by Burnet: "[A] theory which explains everything as a form of a single substance is clearly bound to regard all differences as quantitative. The only way to save the unity of the primary substance is to say that all diversities are due to the presence of more or less of it in a given space." (1920, p. 74). Cf. also Aristotle: "The primary masses, according to [Leucippus and Democritus], are infinite in number and indivisible in mass: one cannot turn into many nor many into one; and all things are generated by their combination and involution. Now this view in a

istic conception, took a quantitative view of empirical reality, that is, became concerned with its *measurable* aspects.[19] Not only was the Aristotelian dictum of focusing attention on the qualitative abandoned, but so was the Aristotelian view that essences are not hidden. As early as with Galileo, Boyle and Newton, atomistic theories intended to account for the nature of heat, gases and light respectively were put forward. Interest had become clearly directed towards the modern-scientific realist aim of gaining quantitative knowledge of the world, and understanding it in terms of efficient causes operating on a deeper epistemological level.[20] And not only had the biologically oriented thought of Aristotle been left behind, but so, in effect, had the doctrines of the Christian Church. Modern science was hereafter able to develop in its accumulation of empirical laws to be explained by theories formulated in terms of three coherent physicalistically interpreted[21] principles – principles not obtained by empirical generalisation, but taken from the metaphysics of atomism.

5. INTELLECTUAL AND PRACTICAL SUCCESSES AND PROBLEMS

The development of the physicalist line in the form of modern science has, on both the empirical and theoretical levels, been more fruitful than the development of any other epistemology. Given its

---

sense makes things out to be numbers or composed of numbers." (*On the Heavens*, 303ª4–10). Cf. also p. 116, n. 8, above.

[19] "With Galilei and Newton the mathematical and functional theories of nature were replaced by a physical theory of nature which was made explicit in terms of the near-at-hand." Northrop (1931), pp. 36–37. The common conception of the influence of earlier thought on Galileo is to place him in the same camp as Kepler and call him a Platonist. This view however cannot explain Galileo's direct espousal of atomism, nor the fact that his major scientific contribution was in keeping with atomism and concerned dynamics rather than kinematics.

[20] Burtt again: "From being a realm of substances in qualitative and teleological relations the world of nature had definitely become a realm of bodies moving mechanically in space and time." (1924, p. 161).

[21] "From Thales, Anaximander and Democritus, via the work of Kepler, Galileo, Newton, Dalton, Fresnel, Faraday, Maxwell, Darwin, Boltzmann and Planck to Einstein, Schrödinger, Watson, Crick, Salem, Weinberg and Gell-Mann, there is the gradual clarification and development of one basic idea, physicalism." Maxwell (1984), p. 238.

core principles, with varying degrees of qualification, its intellectual successes have been many. It has indicated the approximate age of the earth, and revealed that it is physically like other planets, while the sun is like the stars; and it has indicated the rules according to which large and small objects move through space under the influence of such forces as that of gravitation. It has shown what the pure chemical elements are, and indicated the nature of their constitution; it has explained the phenomenon of heat; and it has shown light to be but part of a much wider spectrum of electromagnetic radiation. It has also indicated how we humans are genetically evolved from simpler forms of life, and has revealed the nature of the mechanism of this evolution, as well as explaining many of the processes occurring in the human body. And it has made it possible to determine that there is no intelligent physical life, nor apparently physical life of any form, elsewhere in the solar system.

The practical successes of modern science are also generally thought to be great, those that perhaps first come to mind being in the realm of medicine. Here however one must ask what should constitute an overall improvement in the situation of humankind, and then consider whether modern science has actually led to such an improvement. As regards medicine, for example, it has been claimed that thanks to the influence of modern science the average life-expectancy of a large portion of the world's human population is greater now than it was in the past. On the other hand, however, we are today witnessing mass starvation in Africa which can in part be linked to the decrease in infant mortality due to the employment of modern medical techniques.[22] One way of expressing the problem is that humans' ability to handle social organisation has not developed apace with their ability to gain control over physical nature, so that what in another context might have constituted a clear instance of practical progress does not do so given the present state of the world.

---

[22] Cf. Ellul: "History shows that every technical application from its beginnings presents certain unforeseeable secondary effects which are much more disastrous than the lack of the technique would have been. ... There is scarcely need to recall that universal famine, the most serious danger known to humanity, is caused by the advance of certain medical techniques." (1964, pp. 105, 109).

Another area in which one might be inclined to think of practical progress made possible by modern science is in the area of communication. Computers, telecommunication and aircraft are all dependent for their existence on the developments of modern science. Whether these innovations constitute an improvement in a broader context however may also be questioned. The issue is ultimately a qualitative one, but it may be wondered whether a working day spent in front of a computer screen is an improvement over one spent out of doors, or whether it would not be better that one's friends lived nearby so that they could be visited rather than telephoned, or whether tropical fruits, once one has become accustomed to them, taste so much better than apples and berries. Suffice it to say that what may be considered the practical successes of modern science are more questionable than are its intellectual ones.

But modern science faced a fundamental intellectual problem right at its very beginning with its failure to indicate a mechanism for gravity – a problem it has still been unable to solve not only as regards gravity but all four of the fundamental forces. Furthermore, it is generally agreed that the intellectual successes of modern science have been greater in the past than at present. Newtonian mechanics and the Newtonian theory of gravitation (in spite of the latter's problem of action at a distance), and the Darwinian principle of evolution, might be considered the two greatest achievements of modern science, and the more recent of these occurred more than one hundred years ago. The nineteenth century has been called the century of science, being that in which its core principles were accepted with a minimum of qualification, and in which in physics the basis of thermodynamics and the electromagnetic theory were set forth. In the twentieth century, however, the core discipline of physics had to modify its fundamental principles to an ever greater extent, and in ways which are not altogether compatible with what can be accepted from a common-sense point of view. Both the theory of relativity and quantum mechanics involve ways of thinking which are counter-intuitive in this respect. And if we look at the nearer past we see that, in spite of the great amounts of time and energy applied to research in physics, little progress has been made there in the last fifty years[23] – though this has not been the case in biology.

As regards the practical problems to which modern science has given rise, the first that may occur to one is that of the possibility of nuclear war. But even if such an eventuality were never to come to pass, the problem of dealing with waste materials from the production of nuclear power is one that may haunt the world for thousands of years to come. And we have yet to see what genetic manipulation can have as a result if it should get out of control. Rather generally we can say that modern science with its physicalistic orientation is increasing the rate at which humans can pollute the earth and drain it of its resources, while at the same time failing to provide a conception of reality in which the value of human and other forms of life has meaning.

6. WHAT NEXT?

One can wonder with regard to the intellectual whether modern science has a great deal more to offer to our understanding of the world. Considering its essentially physicalistic orientation, it has succeeded surprisingly well in its applications to biology, explaining virtually the whole of the physical aspect of life on earth without reference to teleological notions.[24] But the nature of life itself, and how it differs from non-life or the physical, lies in principle beyond what the physicalistic categories of modern science could ever be able to handle. And even in the physics of the twentieth century we see clear limitations to the application of modern science's core principles. Some have seen in this a movement towards the religious or the mystical in the heart of physics – but in this way the mystical could be read into every situation which we otherwise fail to under-

---

[23] As expressed by George Gamow: "But, after the thirty fat years in the beginning of the present century, we are now dragging through the lean and infertile years, and looking for better luck in the years to come." (1966), p. 161.

[24] By which is to be understood 'teleological notions of the Aristotelian type.' From one point of view the three principles of modern science can themselves be seen as being teleological, though in a rather weak sense. In this context it may be kept in mind that the origins of these principles pre-date Aristotle.

stand. Another view is that we have come to a limit beyond which chaos rather than uniformity applies to processes in the world – but then it turns out that, as indicated in the previous chapter, such a chaos itself presupposes a uniformity. If a new way lies ahead for human intellectual development, it will have to take as its starting point a fundamentally new metaphysics.

But looking at the state of the world at present and what that state implies about the future, it is not clear how such a metaphysics, or methodology which presupposes it, could ever have the chance to develop. The face of the world is undergoing a physical and biological revolution of a magnitude and a speed it has never experienced before, a revolution which bodes ill for the future of human life on earth. Largely due to the availability and subsequent use of huge quantities of fossil fuel, the population of the world has increased almost fourfold in the last 100 years, and is still growing. When the availability of such fuels begins to diminish – and in the case of oil this will be in the near future – and when the results of the pollution the use of these fuels has caused becomes more severely felt – e.g. as a result of the greenhouse effect and depletion of the ozone layer – humankind will suffer the greatest blow in the hundred thousand years of its existence.[25]

The only way this blow can be lightened – for it would be un- duly naive to think it can be averted – is by drastically reducing the rate of human consumption of both energy and raw materials. But it would seem that for this to become a reality would demand a funda- mental shift in worldview on the part of the majority of the world's population. If we consider the dominant secular and materialistic worldview of the present, it is clear that it has been conditioned by the physicalistic metaphysics of modern science. Could a new meta- physics succeed in laying the foundation for a new worldview; and if so what would be the fundamental characteristics of such a meta- physics?[26]

One aspect that there seems general agreement regarding, among those who have given consideration to the matter, is that the new

---

[25] A more detailed argument supporting this conclusion is presented in Appendix I.
[26] In this regard, cf. further Dilworth (2001).

metaphysics, like the biologically oriented, would have to give a central place to the notion of the whole, while at the same time recognising the success of atomism in its concentration on the importance of the part. One way of expressing this is to say that such a metaphysics would have to maintain a balance between the analytic and synthetic in the conception it provides of reality. And it would have to be capable of applying this distinction not only to some sub-part of reality, such as the physical, but to the whole of it, so that not only life but consciousness, self-consciousness and the unconscious would have a place. In this way it might not only provide the basis for a worldview which promotes a sound relationship between humans and the physical and biological world in which they live, but also the conceptual tools for learning more about those aspects of reality left untouched by modern science.

THE VICIOUS CIRCLE PRINCIPLE OF
HUMAN DEVELOPMENT

In this appendix I shall briefly present a theory of human development.[1] This theory is based on what I call *the vicious circle principle*, which itself presupposes Darwin's principle of evolution, as well as the other principles presupposed by that principle, such as the entropy principle and the principle of the conservation of matter.

As regards humankind, where Darwin's theory of natural selection, based on the principle of evolution, is to explain how we humans came to be, the present theory is intended to explain *how we developed afterwards*. This explanation, in keeping with the treatment of explanation in this volume, will consist in indicating in one coherent conceptual scheme the causal relations that are to be responsible for the main thrust of human development to date. Thus where according to the principle of evolution the key cause of biological evolution is *species'* tendency to *mutate*, on the vicious circle principle the key cause of human development is *humans'* tendency to *innovate*. Further, in keeping with the fact that the vicious circle principle presupposes the principle of evolution, the present theory may be seen as an extension of Darwin's theory in such a way as to explain the particular development of humankind.

Where the principle of evolution is the core principle of biology and is presupposed by all the life sciences, the vicious circle principle is here being advanced as the core principle of *human ecology*, to be presupposed by all the *social* sciences, and thereby to constitute the background against which social structure and change are to be understood.

---

[1] A much more detailed presentation will be found in my book, *Too Smart for Our Own Good*, presently in preparation; one of the first presentations of the vicious circle principle was in my (1994).

1. THE VICIOUS CIRCLE PRINCIPLE

The vicious circle principle (VCP) is both easy to understand and in keeping not only with modern science, but also with common sense. Briefly put, it says that in the case of humans *the experience of need, resulting e.g. from changed environmental conditions, sometimes leads to technological innovation, which becomes widely employed, allowing more to be taken from the environment, thereby promoting population growth, which leads back to a situation of need.* Or, seeing as it is a matter of a *circle*, it could for example be expressed as: *increasing population size leads to technological innovation, which allows more to be taken from the environment, thereby promoting further population growth*; or as: *technological innovation allows more to be taken from the environment, the increase promoting population growth, which in turn creates a demand for further technological innovation.*

Here is the principle in greater detail:

A situation of *scarcity* leads to the experience of *need*
which creates a demand for *new technology*
which in certain cases is *developed* and then *widely employed*
which allows the exploitation of previously *inaccessible resources* –
   renewable, non-renewable or both
the taking of resources, including sources of energy, *reducing the quantity* remaining
and leading to an *increase in polluting waste*
and the *extinctions* of various species of plants and animals
while at the same time allowing an *increase in resource consumption*
and typically the production of a *surplus* of goods
which are normally or often of *lower quality* than those they are replacing
while the availability of the surplus weakens *internal population checks*
allowing *population growth*
and underlying *migration*, first to areas where the new technology is being used to produce the surplus, resulting in *centralisation* and *urbanisation*, then, when possible, to areas where it is *not*, taking it along

the new technology invariably requiring *specialisation* for its use

which gives rise to an *increase* in the *complexity of society* as a
whole

thereby promoting *social stratification*

in which there is an increase in the *property* and thereby *power* of the
upper strata

while the lower strata experience an *increase in work*, and a *worsening*
of their *quality of life*

while the surplus in the hands of the upper strata leads to conflict in
the form of *war* amongst themselves

at the same time as it allows their consumption of *luxury goods*,
many of which can be produced thanks to technological development

as well as providing them with *leisure*

some of which is devoted to *cultural development*: the arts, architecture, philosophy, science

while the presence of the surplus also leads to *increased trade*, i.e.
economic growth, amongst the upper strata

which (together with other things) has the effect of *reducing* the *self-
sufficiency* and thereby the *security* of society as a whole

while the population grows so as to *overshoot* the surplus, i.e. to
over-exploit its resources such that the surplus begins to dwindle

the excess population combined with the reduction in available resources leading to *economic decline*

eventuating once again in *scarcity* and *need* (and possible *population
reduction*).

So there you have the vicious circle principle in greater detail. But
the above still constitutes only a sketch, the central points of which I
shall now fill out. (Note that the present section constitutes primarily
a *presentation* of the VCP; the bulk of its support comes in the next
section.)

Scarcity and Need

All animal species need oxygen, water, food, breeding sites, etc.
Let us call these their *basic* needs. In the case of humans in particular there also exist what may be termed *perceived* needs, such as a

businessman's perceived need to make as large a profit as possible, or a woman's perceived need for a new kind of cosmetic.

When a basic or perceived need is difficult or impossible to meet, it constitutes an *experienced* need. Thus experienced needs may or may not threaten the lives of individuals or their ability to procreate, i.e. be basic needs. In all cases, however, experienced needs result from a *scarcity* of whatever it is that is needed. So, for example, it may be the case that vegetable foods have become relatively scarce, such that the human population has an experienced need for more food – an experienced need which is also a basic need.

In the case of other species, such an experienced basic need is typically brought on by changes the populations of the species have not themselves influenced, such as climatic changes leading to a decrease in rainfall. This can also happen in the case of humans. But what is typical for humans is that such changes are brought about by humans themselves, through the operation of the VCP.

As regards other species in a similar situation, basic experienced need, if prolonged, would lead to a reduction in the size of the population, or to the species' extinction and/or mutation to another life-form in which the need is not experienced or is less severely felt. In the case of humans, on the other hand – and this is a key aspect of the VCP – the experienced need may be overcome via *technological innovation*.

Innovation

As suggested by a number of authors, including Thomas Malthus, Ester Boserup and Richard G. Wilkinson, in situations of scarcity, necessity can become the mother of invention. This does not mean that every instance of experienced need will lead to invention – all that is required for the vicious circle to operate is that *every now and then* a technological solution be found. Though the discovery of such solutions may be rare, once a useful innovation is hit upon it is remembered, and its use spreads to other cultures, being easily transmitted largely thanks to the objective nature of technology. Knowledge of how to employ a particular innovation thus spreads to all areas of the world where it can be of use, i.e. where its employment relieves experienced need, and its introduction is followed by migration to those places.

Resource Depletion

The use to which the new technology is put typically involves increasing the amount taken from the environment by making available resources that were previously *inaccessible*. This can mean either allowing direct access to the resource in question, as in the case of mining, or transforming conditions so that what was previously not a usable resource becomes one, such as in the case of land-use changes or petroleum refining. Often it will involve the partial substitution of one resource for another in such a way that what is produced from the two resources together is greater than what was obtained from the original alone. Garden produce may be consumed along with naturally growing plants; synthetic materials made from oil may be supplemented to and partially substitute for those made of cotton, wool and leather. Though efforts to extract the original resource may in principle be reduced when the technology permitting the extraction of the substitute is in place and functioning, this seldom happens in practice; whether it does or not, and whether or not either of the resources is renewable, the result has consistently been a decrease in the total resources remaining.

Thus when new technology is employed to extract more from the environment, the result is a *reduction* in the quantity of remaining resources, and a *permanent* such reduction if the resource in question is non-renewable. Here it is important to distinguish resources from reserves and stocks. *Resources* are a part of nature that, given present and future technological development, could be of immediate use to humans. *Reserves* are those resources which are available given present technology. And *stocks* are those reserves already taken from nature and set aside for future use.

Though we can be sure that the quantity of non-renewable resources is constantly diminishing, we cannot be sure exactly what our future reserves will be, since we do not know whether technological innovations in the future may increase accessible resources (reserves) or reveal resources which we today do not see as such. So, for example, uranium, always a resource, was not appreciated as such – not turned into a reserve – until the advent of nuclear technology.

Thus, with the implementation of a new technology, both reserves and stocks may be increased, and it is the quantity of these, not resources, that directly affects the economic value of such substances. So the implementation of a new oil-drilling technology, for example, may increase oil reserves and lower the price of oil, while at the same time the quantity of oil as a resource is constantly diminishing.

Here the distinction between renewable and non-renewable resources is important. What has made the vicious circle particularly vicious is its undifferentiated involvement of both kinds of resource, and the human dependence on non-renewable resources that has resulted. On the other hand, no resources are renewable if they are overexploited. While populations of large mammals can constitute a renewable food resource for humans, if the species they represent is driven to extinction – as many were during the Upper Palaeolithic – they are not only non-renewable but become non-existent.

To this it may be added that in general humans' increasing extraction of resources (and the resultant increase in waste their use gives rise to) leads to increasing numbers of extinctions of other species – both plants and animals. Thus not only may such species be eradicated directly, e.g. by over-hunting, but also indirectly through the diminution or spoiling of their resource base.

If all potentially renewable resources were renewed or allowed to renew themselves, as is normal in the case of other species, and if at the same time we had not become dependent on the use of non-renewables, then it would have been possible to avoid getting caught up in the vicious circle, as many modern hunter-gatherer communities apparently had not before the intrusion of other cultures. But dependence on non-renewable resources leads to a situation in which technological innovation becomes a must in order to extract a replacement when the resource in question has been exhausted, just as it becomes a must that there exist such a replacement to turn to.

Waste and Pollution

If the energy used to extract the newly available resources is biotically based, and if those resources themselves are biotic, waste will be decomposable. But if non-renewable fuel sources are used to produce non-biodegradable products, pollution and accumulating garbage will result. Though some non-biodegradable products may be recyclable, the second law of thermodynamics makes perfect recycling impossible, so all such products eventually become waste.

Surplus

In the case of e.g. a scarcity of vegetable food, at one particular stage of human development technological innovation consisted in the development of the lance, which allowed an increase in the consumption of meat as a replacement for vegetable matter. At another time it took the form of irrigation, allowing the growing of crops in areas that would otherwise have been too arid. Once the use of the new technology becomes widespread, the result has often meant a shift from a scarcity to a *surplus*.

Note that the surplus, though meeting the same need, may be of a resource different from that which was scarce, and that it can be quite great. What is typically the case, and is very clear in the development of the lance, is that the resource that was scarce (vegetable food) is at least partly *replaced* by another resource (meat) that meets the same experienced need (more food). Other examples of this phenomenon include the replacement of meat with vegetable products thanks to the technological innovations of the horticultural and agrarian revolutions, the replacement of wood with coal at the time of the industrial revolution, and the replacement of whale oil with petroleum, beginning in the middle of the nineteenth century.

In the case of the development of the lance, the new food source, meat, was superior (higher in protein content) to what it was replacing. However this is not always the case, and, as has been implied by Wilkinson, more often has *not* been the case, such as when meat was largely replaced with vegetable products at the time of the horticultural revolution. Though the use of the digging stick – and later the hoe – in planting tubers allowed the harvesting of much *more* food than was available before, the food was of a lower quality.

What may be noted here, particularly against the background of the idea that necessity is the mother of invention, is that, as argued by Wilkinson, after an innovation has been made, necessity has more been the mother of the *employment* of invention. Here the VCP, in saying that invention is paradigmatically incumbent upon experienced need, supports Wilkinson's view that technological change is typically *implemented* in a situation of need, as against the view that such change is the result of the seeking after a better life.

Social Stratification

Social stratification, or a 'pecking order,' is a manifestation of Darwinian intraspecies survival of the fittest, and exists in the populations of virtually all medium- to large-size animals. In the case of humans, technological development has made possible a pecking order that has now become global, where few have much power and many have little or none. The invention of *language* (which itself may be seen as a tool) is important here, for it is through language that orders can be given and passed on to those not present. *Weapons* are also important, in that military strength is the bottom line when it comes to power.

Pushed by the VCP, the destructive capability of weapons has constantly increased throughout our development, and channels of communication have constantly grown. This has had the effect of steadily increasing the power of the strong and lessening that of the weak, at the same time as it increases the complexity and reduces the security of society as a whole.

This development, involving weapon and communication improvement, globalisation, centralisation, and population growth, is a manifestation of the VCP, the operation of which works counter to the survival of the human species, involving as it does constantly increasing consumption, population size and quantities of waste, all of which tend to move the species out of equilibrium with its surroundings, thereby increasing the likelihood of its becoming extinct.

Quality of Life

With social stratification come of course inequities. While the quantity of resources going to the weak is normally only sufficient to allow them to raise children, that which remains – i.e. virtually the whole of the surplus – goes to the powerful. Thus, given sufficient surplus, there will develop among other things the production of *luxury* goods over and above the goods necessary for the survival of the population. Such goods may themselves be of a technical nature, as are e.g. stereos and pleasure boats, and thereby constitute instances of non-typical technological change, i.e. technological change that does not support the maintenance or growth of the population, but merely fills non-basic perceived needs.

Leisure, or the potential for leisure, will also increase for the powerful. The efforts required to produce usable products from natural resources will be those of the weak, the greater the disparity between the powerful and the weak, the less the leisure for the latter.

Thus, so long as there exists sufficient surplus, it is to the advantage of the powerful that the weak bear many offspring, to do their work for them and to fight in their armies. Population growth amongst the weak widens the gulf between them and the powerful through reducing the value of an individual's labour; and it increases the military strength of the powerful vis-à-vis others in power by providing them with more soldiers – the existence of both factors perhaps explaining why those in power have constantly turned their backs on problems of overpopulation.

The fact that a particular resource can only be significantly exploited thanks to the introduction of new technology implies that the acquisition of that resource should require more work than did that of the original resource before it became scarce. Thus, following Boserup and others, we see that the employment of horticultural technology in the Neolithic era, involving e.g. the construction and use of stone gardening implements, requires more work per unit food acquired than did the earlier hunter-gatherer technology. This line of reasoning takes a slightly different form when applied in a modern context, for much of the extra work required in modern agriculture and other production is done by machines, which obtain their energy from fossil fuels. So we should then say more generally that the use of new technology tends to require the expenditure of more *energy* than did the technology it is replacing. Still, with the first introduction of a new technology, before the supporting non-human infrastructure has been built up, an increase in the amount of work performed by the weak has been the rule even in the industrial era.

That the weak have less leisure and do more work means a lowering in the quality of their lives, particularly for those drawn into the extraction of resources or the production of the goods or services resulting from the implementation of the new technology. The more onerous lifestyle for the weak that accompanies the transition to using a new technology, and the generally lower quality of their food in particular, raises their mortality. However, as the use of the new technology becomes an integral part of society, this effect will tend to lessen – until the next more effective (exploitative) technology is implemented.

Mortality from infectious diseases is another aspect of the VCP. We have acquired a vast number of our infectious diseases from the animals we domesticated in our technological development of horticulture. In conjunction with this, the constant growth of the human population steadily increases the breeding grounds of the bacteria and viruses responsible for these diseases, thereby constantly increasing the likelihood of major epidemics.

Conflict

On the VCP, the killing of humans by other humans is a population check that expresses itself particularly in situations of overcrowding, and has been made possible or greatly enhanced through the development of weapon technology. In our primitive past, human males, like the males of other species, would fight with each other over breeding territory and/or females, the losers often having to go without a mate. But as it became possible for ever fewer people to accumulate ever-increasing quantities of wealth and military power, this fight over territory has taken the form of *wars* between *states*, i.e. between the *leaders* of states. War, among other things, is a check to the size of the total population manifest in the fighting of powerful individuals over territory.

Cultural Development

The leisure had by the powerful, in combination with the human tendency to innovate, has given rise to the arts and philosophy. And in combination with constantly increasing technological know-how, it has allowed for their development e.g. in the form of monumental architecture and science. Thus, on the VCP, with its strong biological-ecological orientation, the arts and sciences are an 'emergent property' of the basic dynamics of human development – they could be seen as a side-effect. In the case of modern science, however, as expected already by Francis Bacon, some of the results of the search for knowledge and understanding of the physical world have been channelled back into the productive effort (particularly that of the *military*), thereby speeding up our course round the vicious circle.

Economic Growth

As suggested above, the implementation of a new technology will quite generally mean an increase in society's use of *energy*. Historically, the first non-human source of such energy was wood used in fires, and later domesticated animals such as the ox and the horse, and since the industrial revolution mainly fossil fuels. Note that while energy is itself a non-renewable resource (though its *source* may be renewable) which in certain cases can be made available via technological change, unlike metals and certain other resources, it cannot be recycled.

All use of technology demands energy, generally the more sophisticated the technology, the greater the energy required per unit it produces. This extends to trade, such that the energy required to trade a particular entity increases with the distance between the points of trade. Increase in the availability of energy thus promotes economic expansion, thanks both to the increasing number of products technological development makes available, and the ease of transportation it makes possible.

Thus the surplus resulting from the use of the new technology may be, and in modern times virtually always is, put on a market, thereby giving rise to trade, or, in the case of ongoing trade, giving rise to an *increase in trade*, that is, to *economic growth*. If the surplus is sufficiently great, this economic growth will also be manifest in related areas, such as transportation, involving e.g. an increase in the number of transport vehicles and the improvement or building of roads and other transportation systems. In this way self-sufficiency decreases, and there is a further reduction in the security of the population.

Population Overshoot

The paradigmatic increase in available resources incumbent upon technological change is an increase in *food*. At first such an increase may mean more to eat for each person. But what invariably happens in the longer term is that it leads not to people consuming more, but to *more people consuming*. That is, the surplus resultant upon the technological change, given an increased number of breeding sites (housing) also made available by technological development, will lead to *population growth*. This is another key aspect of the VCP, namely, that technological change employed to counteract need often *overshoots the mark*, giving rise to a surplus and a consequent increase in population. Provided with an improvement in our immediate conditions for survival, we, as would the members of other animal species, generally succeed in seeing to it that more of our offspring live to an age where they themselves can reproduce.

As has been emphasised by Virginia Abernethy, throughout human history periods of surplus have as a matter of fact been followed by periods of population growth. Such seemingly minor innovations as the adjustable wrench can have the ultimate effect of providing or increasing a surplus. The increase in the amount of resources that can be extracted from the environment is then taken as permanent, and what Abernethy terms a 'euphoria effect' takes hold, leading people to have larger families. Note that the increase in average family size can be extremely small and still result in marked population growth.

Not only does population grow when provided with a surplus, but it grows beyond what the surplus – which is itself dwindling partly due to its being used by a constantly increasing population – can support. Eventually the surplus will be eradicated, thereby closing the vicious circle and taking us back to a situation of *scarcity* and experienced *need*.

2. APPLICATION AND CORROBORATION

In order to investigate the viability of our theory of human development based on the VCP, we shall consider the extent to which actual human development to date can be understood in its light. In what is to follow, factual information will first be presented concerning a major period of human development, and then the theory based on the VCP will be applied to that period. Thus the present section is intended both to *provide support* for our theory and, granting its viability, to *explain* the main outlines of human development to date.

The Cro-Magnon (Hunting) Revolution – 100,000 BP

Modern humans originated in Africa some 200,000 years ago, coming first to the Middle East around 100,000 BP, and eventually to Europe about 40,000 BP. At each of these last two times – the first of which was just before the beginning of the last ice age, and the second of which was in the middle of it – there was a marked increase in innovations.

The innovations of 100,000 years ago include the use of bone for tools, the making of tools with built-in handles, and probably the use of skin clothing. The innovations after 40,000 years ago include the systematic hunting of selected animals, the widespread use of blade tools, the ability to create fire, and the invention of lamps, needles (bodkins), spoons, pestles, stone axes, the spearthrower, and, about 23,000 years ago, the bow and arrow.

At each of these times, i.e. 100,000 and 40,000 years ago, there was a spurt in human population growth. More generally, i.e. over the Cro-Magnon era as a whole, what we see is continuous population growth up to the Mesolithic period (12,000 BP).

Perhaps what is most notable with regard to human technological development during the last ice age (80,000 to 10,000 BP) is the improvement in *weapons*, more particularly the development from the wooden lance to the stone-pointed throwing spear, to the throwing spear which could be hurled a greater distance with the aid of a spearthrower, to the spear with a sharpened bone point – later with barbs on it, and on to the bow and arrow, eventually with poison-tipped arrows. It may in fact have been the case that the demise of the Neanderthals about 28,000 years ago was a result of most if not all of this development of weapons occurring with us. To this it may be added that throughout human existence the constant improvement of weapon technology has made itself felt in an absolute increase in the killing of humans by other humans.

By the end of this first period of human development, many species and genera of large mammals over the whole world had become extinct. Africa lost 50 of its large mammalian genera – some 40 per cent – after modern humans came into existence, with the peak and end of extinctions there occurring at around 40,000 BP. Australia lost 21 genera of large mammals (86%) after the arrival of humans, peaking at about the same time; and in North America, during the 2000 years after a wave of humans arrived around 12,000 BP, 31 genera (60%) of the mammalian megafauna became extinct, with a subsequent similar reduction in South America.

In North America, for example, the extinct animals include the American mastodon, the Colombian, woolly, Jefferson's, imperial and pygmy mammoths, the tapir, the stag moose, two species of deer including the giant deer, the plains camel, the llama, the native American horse, the large long-horned bison as well as another subspecies of large bison, a species of pronghorn, the Asian antelope, the yak, four species of ox including the shrub-ox and the woodland musk-ox, the sabre-toothed and scimitar-toothed cats, the cheetah, the spectacled and giant short-faced bears, the dire wolf (a giant wolf), the giant beaver (the size of a black bear), two species of peccary (related to swine), the anteater, the glyptodon (one of the largest ancient armadillos), and various ground sloths including the Shasta (the smallest, but still the size of a black bear), and the giant (the size of an elephant); and non-mammalian extinctions include that of the giant tortoise.[2]

After such decreases in food resources, i.e. at the beginning of the Mesolithic era and a couple of thousand years before the horticultural revolution, humans generally became more omnivorous in their diet, killing smaller animals and eating less meat. Their stature decreased by about five centimetres, and the size of their population began to diminish, at least in Europe.[3]

VCP Explanation of the Cro-Magnon Era

On the VCP, the spurt in innovativeness about 100,000 years ago was the result of many factors. One was of course *Homo sapiens'* greater intelligence than other human species. Another factor must have been our movement into a new area *north* of where we came from. Thus we did not develop skin clothing in tropical Africa 150,000 years ago, but in the Middle East 100,000 years ago; and we did not learn to create fire until some 40,000 years ago, just prior to our entering Europe. Biologically we are tropical animals, but thanks to our technological innovativeness we have been able to adapt to harsher climates. And this adaptation has of course allowed our population to grow, the acceleration in population growth both 100,000 and 40,000 years ago being partly the result of our colonising new areas.

On the VCP, it was thanks to our technological creative ability that the migrations during our Cro-Magnon stage were possible. This migration was *prompted* by an increase in our numbers also thanks to that ability, and *resulted* in a further increase in our numbers. But this momentum of population growth ran into a wall when we no longer had new areas to move to at the same time as, again thanks to our technological ability, we had eradicated most of our large game.[4]

---

[2] Martin (1966), pp. 339 and 340; (1967), pp. 79 and 111; (1973), p. 969, 972–973; (1984), p. 170; Martin & Guilday (1967), p. 20; Baerreis (1980), p. 356; Barnowsky (1989), p. 236; Ponting (1991), p. 35.

[3] Cohen (1989), p. 112; and (1977), p. 126.

[4] Ibid., pp. 14–15.

As regards the demise of large game animals and many of their predators, we see the VCP at work in a very clear form. The technological innovations that allowed humans to expand into new areas provided them with a *surplus* of food, which allowed more people to be fed and led to accelerated *population growth*. This population growth was a contributing factor to the *over-exploitation* of our food *resources*, manifest in the *extinctions* of those species that were the most accessible. With their demise there arose a situation of experienced *need*. And this persistent need, exacerbated by continuing population growth, led to *technological innovation* in the form of more sophisticated weapons that could be used against smaller and more elusive prey.

The Horticultural (Domestication) Revolution – 10,000 BP

Starting around 10,000 years ago, humans began changing their lifestyle from that of hunting and gathering to herding and primitive agriculture. Herding of course took place where there were animals to be herded, and such areas were naturally more arid than those in which actual horticulture was employed.

What perhaps best characterises this time is the move to domestication. Herding is the first step towards the domestication of animals, and the horticultural lifestyle itself involves the domestication of both plants and animals.

As regards plant domestication, the first and subsequently predominant form of horticulture was swidden, or slash-and-burn, cultivation. First, an area of forest was cut down (using newly invented polished-stone axes) and then burned, after which women stuck tubers in holes they made in the area using digging sticks. Each year the productivity of the particular plot would decline, until it was preferable to chop down a new area of forest and start fresh. The original patch would be left fallow to eventually regain its forest character, after which it could again be used for crops.

At this time population growth began to accelerate once again. Despite the fact that human life-expectancy fell from about 30 years for hunter-gatherers[5] to about 20 during the whole of the horticul-

---

[5] Angel (1975), pp. 182–183.

tural era,[6] during this period the human population increased from some 5–10 million to 80 million.

With increasing population density, however, the length of time a swidden plot would be used before being left fallow increased, and the length of time it was fallow shortened. Longer use of the plot increased the amount of weeds, and the shorter fallow period did not allow the regrowth of forest but only bush, also resulting in increased weeds. This increase in weeds was responsible for the transition within horticulture from the use of the *digging-stick* in *forest-fallowing* to the *hoe* in *bush-fallowing*. Bush-fallowing is physically much more demanding than forest-fallowing, taking a longer time and involving the use of a heavier tool. More work was also required to grind the stone heads of the hoes and other implements needed for this form of horticulture. Eventually bush-fallowing was in many places replaced by *gardening*, where the same plot was used continuously, and crop-rotation was employed to support productivity. This specialised food production meant a lack of flexibility which could lead to starvation when crops failed e.g. as a result of drought.

As mentioned, the horticultural era also involved the original domestication of animals. This began with the dog (from the wolf) some 12,000 years ago or earlier, and was succeeded by goats and sheep around 9000 years ago, then cattle, pigs and bees, and finally the horse and donkey about 6000 years ago. In all these cases the domestic animals were smaller than their wild predecessors.

The domestication of animals led to humans' contraction of many diseases. These include smallpox, brucellosis, malignant boils (anthrax) and tuberculosis from cattle, influenza from pigs, leprosy from water buffalo, the common cold from horses, and measles, rabies and tapeworm from dogs.[7] All told, we now share 65 diseases with dogs, 50 with cattle, 46 with sheep and goats, and 42 with pigs.[8] The effect of these diseases constantly increased with the increase in population density due to population growth, and with constantly increasing trading; and our sedentary lifestyle led to more unsanitary

---

[6] Clark (1989), pp. 84–85.

[7] Cf. McMichael (1993), p. 91.

[8] Ponting (1991), p. 226.

conditions which further supported the spread of parasites (including diseases) generally.

Warfare was virtually impossible for hunter-gatherers, not only due to their small numbers, but also because of their not being able to accumulate sufficient food to see them through such engagements. (Not that hunter-gatherers did not manage to use their weapons to kill one another anyway, only on a smaller scale.) Thus it was only with the horticultural era, when food could be stored, either as grain or livestock, that real warfare became possible. Furthermore, it was the existence of *property*, any great quantity of which hunter-gatherers lacked, that constituted the immediate cause of war, as well as being a prerequisite for commerce.

Apart from the direct loss of life resulting from such conflicts, warfare also supported female infanticide. By reducing the number of girls in a society, the group could devote its resources to the nurture of its male warriors-to-be. In a survey of studies of 609 horticultural societies it was found that the sex ratio among the young was most imbalanced in those societies in which warfare was current at the time of the study, and most nearly normal in those societies in which warfare had not occurred for more than 25 years. In the former, boys outnumbered girls by a ratio of seven to five on the average, indicating that nearly 30 per cent of the females born in these societies had died as a result of female infanticide or neglect.[9]

During the horticultural period social inequality constantly increased as ever more elaborate systems of social stratification replaced the relative egalitarianism of the past. Slavery was invented; and poverty, crime and war became widespread, while the rate at which humans degraded their environment also increased.

The tools developed during the horticultural era apart from the polished-stone axe and the hoe included sickles, cloth, woven baskets, sailboats, fishnets, fishhooks, ice picks and combs.

VCP Explanation of the Horticultural Era

On the VCP, and in keeping with the anthropological and archaeological findings of the past 50 years or so, the horticultural revolu-

---

[9] Divale (1972), p. 228.

tion was not a change to a better lifestyle made possible by humans' innovative ability, but a change to a worse lifestyle for a relatively and absolutely greater number of people made possible by that same ability. Rather than suffer great dieback when we had over-exploited our food resources – as other species would have – we found a way of staving off such an eventuality through the widespread adoption of horticultural technology.

What this change meant, however, among other things, was a general lowering of the quality of our lives. This was manifest in increased work, a poorer diet, increased killing of other humans through war and infanticide, and a tremendous increase in infectious disease and parasites. Poorer diet and increased disease, and quite possibly the kind of work we had to do, led to a further decrease in our stature,[10] and these factors together with increased killing of humans by other humans resulted in a drastic reduction in longevity.

The VCP is constantly operative, being responsible for the steady increase in work required as we passed through forest-fallowing to bush-fallowing and on to gardening. Thus, thanks to the VCP, while the amount of work increased, it was still possible to maintain a surplus of food, which in turn supported about a *tenfold* increase in the human population during this period, despite the hardships experienced by individuals.

Population checks, whether internal or external, or positive or preventive, operate so as to *counteract* the VCP. What we see with the expression of the VCP through the whole of human development is a steady *weakening* of these checks at the same time as the constantly altering conditions and increasing complexity of human society lead to their taking new forms, or to certain forms manifesting themselves to a greater or lesser degree than earlier. *Warfare*, and particularly the female *infanticide* that has accompanied it, is a positive internal check that probably began in the horticultural period and has been with us ever since. The population checks that became weakened during this period were primarily those related to the spacing of children. Where for hunter-gatherers there were some four or five years between births, this spacing shortened noticeably

---

[10] Cohen (1989), p. 112, n. 32.

with the horticulturists. On the VCP the primary, though not only, reason for this change was sedentism combined with the virtually constant presence of a *surplus*.

So, quite generally, we see the VCP operating through the horticultural era as follows: First there is a situation of experienced *need* for more food resultant upon *population growth* and the *extinction* of food sources during the hunter-gatherer era. This need leads to the employment of *technological innovations*, the most important of which consists in the domestication of plants and animals. This in turn results in a *surplus* of food of *lower quality* than what it is replacing, the acquisition of which demands more *energy* in the form of human *labour*. This state of affairs means a *lowering* of the *quality* of most people's *lives*, while at the same time the surplus provides the basis from which the *population* is able to continue to *grow*.

The Agrarian (Ploughed-Field) Revolution – 5000 BP

The most important technological development after the beginning of the employment of horticultural methods was the invention and use of the plough. The plough (developed from the hoe), and shortly afterwards irrigation, made possible the maintenance and continued growth of the human population. With the plough, seed crops could be grown more widely, largely replacing the hand-planted tubers of the horticulturists. The resultant harvest per unit land in terms of calories increased tremendously. Worldwide, where conditions were suitable, seed cultivation expanded at the expense of horticulture, while at the same time these cereal crops depleted soils more quickly than did the vegetables and fruits of horticultural agriculture.

Though the average age at death did not change notably in the transition to the agrarian period, remaining at about 20 years, such phenomena as the drought and flooding resultant upon changing weather were more widespread in their effects. Similarly, the spread of infectious diseases had a greater impact due to the constantly increasing size of the population and greater interaction between groups, coupled with worsening hygiene. The most devastating instance was the Black Death in the middle of the fourteenth century, which killed about a third of the population of Europe over a period of four years. And just as the human workload grew with the move to horticulture, it grew again with the move to agriculture. Though oxen and later horses were eventually used to plough fields, at the beginning of the agrarian era it was *humans* who drew the ploughs.

With the amassing of greater amounts of property, society became more stratified: there was a greater division of labour, and the distance between the rich – the richest of whom was the king – and the poor constantly increased. Not only did the lower strata have to do more work, but the top stratum had become more powerful, consisting of relatively fewer persons and having larger territories. The vast majority of the population worked as peasants or slaves for the king's personal best interest: they laboured creating the agricultural surplus that constituted his wealth; they built monolithic monuments to his greatness; and they fought in his wars.

For the last 200,000 years, humans' adaptation to their environment has taken the form of cultural and genetic change rather than chromosomal change, the last chromosomal change being the evolution of modern man. With the development of commerce and warfare much of that cultural and genetic change has consisted in adapting to the activities of other humans rather than to nature. By the time we reached the agrarian period we no longer had to fear other predators, and for the most part had a dependable source of food through the domestication of plants and animals. Thus major successes consisted in the results of exchanges on the market or on the battlefield, rather than in the results of the hunt.

As regards war, the conquest of other people became a profitable alternative to the conquest of nature. Beginning in advanced (metal-using) horticultural societies and continuing in agrarian, almost as much energy was expended on war as in the more basic struggle for existence. At the beginning of the agrarian era, the time of year after the harvest was in was known as 'the season when kings go forth to war.'[11] War was a regular part of life. Here the development of bronze weapons, and later steel weapons, each played a key role.

---

[11] L. Woolley, cited in Lenski et al. (1995), p. 181.

The first kings of cities were originally those lords of villages who had an advantage when it came to such weapons.

Nature gradually became less the 'habitat' of the farmer and more a set of economic resources to be managed and manipulated by those in power.[12] With this continual distancing of nature from humankind came a change in our conception of the world, with reality increasingly being thought of in terms of people and their actions.

While the quality of humans' lives declined, so too did the quality of the land. The removal of trees – eventually with the aid of iron axes – to allow the creation of fields led to soil erosion, a process accelerated by the exposure of the topsoil to wind and rain by the use of the plough. In areas which were too hilly or deficient in nutrients for agriculture, goats were let roam, in effect exterminating all but the hardiest bushes. The results of these activities remain with us today, and can be seen, for example, over the whole Mediterranean area. The use of irrigation to increase or make possible the supply of water to fields led to salinisation, leaving the soil unusable for cultivation for thousands of years – as can be witnessed today in the Tigris-Euphrates valley. Wastelands developed around cities, where the area was picked clean of fuel and building materials, and vegetation disappeared due to pedestrian traffic.

As regards other species, the spread of agriculture and the human population forced many from their natural habitats, and some were driven to extinction. For example, around 200 BC the lion and leopard became extinct in Greece and the coastal areas of Asia Minor, and the trapping of beavers in northern Greece led to their extinction. Somewhat less than 2000 years ago, the elephant, rhinoceros and zebra became extinct in North Africa, the hippopotamus on the lower Nile and the tiger in northern Persia and Mesopotamia. Whales were hunted to extinction in the Mediterranean before the fall of the Roman Empire; the crane became extinct in the 1500s; and the last wild ox was killed in 1627.[13]

As in the horticultural period, in the agrarian era one internal population check was infanticide in the face of the threat of severe deprivation. When crops failed and famine was imminent, families

---

[12] Roberts (1989), p. 128.
[13] Ponting (1991), pp. 161–163 and 187.

often abandoned their new-born by the roadside, or left them at the door of a church or monastery in the hope that somebody else might raise them. Sometimes even older children were abandoned by their parents. The story of Hansel and Gretel is based in the reality of scarcity. In some districts in China as many as a quarter of the female infants were killed at birth – signs were put up near ponds, "Girls are not to be drowned here."[14] Female infanticide at such a level was not exceptional in such cultures, but, in conjunction with annual warfare, was rather the *norm*. Other population checks included setting minimum age and capital requirements for marriage, and the directing of adolescents into religious careers as priests or nuns.

The length of the agrarian period was extended by the discovery of America, which also allowed a continuing increase in the size of the human population through emigration from Europe. At the same time, however, population also increased in Europe, partly due to the taking of resources from the New World.

The technological innovations of the agrarian period included, besides the plough and iron axes, such inventions as the potters' wheel, the wagon wheel, animal harnesses, horseshoes, stirrups, the lathe, the screw, the wheelbarrow, the spinning wheel, printing, and watermills and windmills. From 5000 to 250 years ago the human population increased from about 80 to 790 million, thanks primarily to agricultural innovations, which increased the number of calories available for human consumption while at the same time reducing the productivity of the land.

VCP Explanation of the Agrarian Era

With the beginning of the agrarian era, the vicious circle of population growth, technological innovation, and resource depletion was about to be traversed once again, and on a massive scale. According to the vcp, the agrarian revolution was a result of the same forces as brought about the horticultural revolution. The need for more food on the part of a constantly *growing population* led to the *invention* and *use* of such tools as the plough and the technology of field agri-

---

[14] R. K. Douglas, cited in Lenski et al. (1995), p. 199.

culture, including irrigation. The result was once again a tremendous increase in the quantity of food, giving rise to a *surplus*, which in turn supported a further increase in population. And this increase continued throughout the agrarian era thanks to the continual invention of new technology, allowing *ever-increasing quantities of resources* to be taken from the environment, either directly or indirectly, providing on the whole the maintenance of the surplus. Technological development continued to be, not the result of a seeking after a better life, but the response to experienced need resulting from the presence of increasing numbers of people.

With this increase in production and population came an increase in *resource depletion*, mainly in the form of the leaching and salinisation of agricultural soil, the extermination of some species of animals and a reduction in the populations of others, and an *increase in the production of waste. Social stratification* increased further due to the technological innovations' involving improved communication and weapons; and while the poor continued to live lives of drudgery, *infectious diseases* and the presence of other parasites increased among the powerful and the weak alike – though better nutrition, hygiene and living conditions would have made these effects less severely felt among the powerful.

War increased partly as the manifestation of a check to the ever-growing population, becoming more pervasive and on a larger scale with its growth. Likewise *infanticide* continued, further checking population growth, the employment of both sorts of check being morally obligatory,[15] just as killing in war is still morally obligatory today. But primarily as a result of the magnitude of the agrarian surplus, these checks, in combination with other checks related e.g. to marriage customs, failed to curb the continuing growth of the population.

The Industrial (Fossil-Fuel) Revolution – 250 BP

With the industrial revolution came not only a huge increase in the use of machines, but a similar increase in the use of fossil fuel –

---

[15] A conception of the biological basis of morals in keeping with the VCP is presented in Dilworth (2005).

more particularly coal – to power them. In Britain, around 1750, the growing population, among other things, led to a shortage of both land and wood, the latter being used as a building material and a fuel. The possibility of using coal as a substitute fuel for wood was hindered by water constantly seeping into the coal mines. This problem was overcome through the invention of machines – first the Newcomen engine, and later Watt's steam engine – to pump the water out of the mines.

Coal is an inferior substitute for wood as a fuel, a fact which required the invention of other devices and processes such as the use of coke (derived from coal) as a substitute for wood-charcoal in the iron-smelting process – the over-use of charcoal having been a major cause of the dearth of wood. With these changes and the implementation of Watt's engine, the use of coal increased tremendously.

Since coal only existed in particular places, it was necessary that a means of distribution be found. This means involved the invention and use of steamships and locomotives, the construction of canals and railways, and the development of hard surfaces for roads.

The industrial revolution also brought with it a worsened situation for labourers. Instead of working out of doors or in cottages and enjoying the many holidays of the agrarian year, people worked in mines or factories for longer hours at more monotonous tasks and with virtually no holidays. Unlike in the case of farming, mechanised work can be done throughout the year, and with artificial lighting even at night. Child labour also increased, with children being used, for example, in mines, where they could squeeze into spaces too small for adults. The quality of clothing also became lower, with cotton largely replacing linen, wool and leather.

Migration increased from the less-developed to the more-developed areas, from the country to the city. And afterwards, as improved transportation opened up the colonies, there was a massive migration to the Americas and Australia.

In the United States the transition to steam power and the use of coal as a fuel did not take place until there were local wood scarcities and shortages of sources of water power. Quite generally, Americans used less-sophisticated technology when they could, even though they were familiar with the alternatives. They clothed themselves in furs and leather rather than woven cloth; they used wood rather than coal, water power rather than steam, and herding (ranching) and primitive extensive agricultural methods rather than the more intensive European methods they were already familiar with.[16]

Since the industrial revolution *wars* have become larger and more destructive, with larger armies and more powerful weapons, the latest on the list of weapons being nuclear devices. The twentieth century involved greater destruction and loss of life from war than any previous 100-year period.

Trade has also increased tremendously during the industrial era, and is still growing. The increasing production of goods incumbent upon technological innovation and the use of fossil fuels has been disbursed to markets near and far thanks to such innovations as fuel-powered ships and aircraft.

As mentioned, the technology of the industrial era has been operated using huge quantities of fossil fuels, the second of major importance being *oil*. It is the use of these non-renewable fuels more than anything else that perhaps best characterises the industrial era. Where coal began to be used when wood became scarce, petroleum began to be used in response to a scarcity of whale oil. At present, 77 per cent of the energy used in the world comes from fossil fuels, about 33 per cent from oil (the extraction of which will peak in the near future), 26 per cent from coal, and 18 per cent from natural gas.[17] And the rate of use of fossil fuels is almost five times what it was 50 years ago, in spite of international recognition of the many environmental evils to which their use gives rise.

Furthermore, oil is used in the making of a myriad of synthetic products, including most forms of plastic and an ever-increasing proportion of textiles, both of which are invariably of inferior quality to what they are replacing. In the case of textiles, just as cotton became a substitute for linen, wool, and leather during the industrial revolution, we are today witnessing the substitution of oil-based synthetic products for cotton. And where the members of a family could once clothe themselves in woollens and leather using their own raw materials, we are now dependent for clothing on oil wells,

---

[16] Cf. Wilkinson (1973), pp. 171–172.
[17] von Weizsäcker et al. (1997), p. 251.

oil tankers, refineries, the chemical industry, textile-machinery man-
ufacturers, and the metal and power industries needed to back them
up. The same goes for our food, heating, transportation and almost
every other item we consume.[18]

The increase in the production of goods during the industrial period has also meant an increase in waste products. These include not
only the goods themselves when they are no longer of use, but the
pollution resulting from the burning of the fossil fuels used in their
manufacture and operation. And where the animal domestication of
the horticultural revolution led to a tremendous increase in infectious
diseases among humans, the pollution resultant upon industrialisation
and the chemicals used in production have led to a similar increase
in cancer.

Extinctions of other species either directly or indirectly due to the
activities of humans continued into the industrial era. For example, the
osprey, which was common in the 1700s, was hunted to extinction
in the 1800s. The great bustard became extinct by 1838. The last
pair of great auk, a flightless seabird, was killed in Iceland in 1844.
The sea eagle was still common as late as the 1870s, but is now
extinct. The Bali tiger became extinct in the 1940s; the Caspian tiger
in the 1970s.[19] The human-related wave of extinctions that began in
Palaeolithic times and has been accelerating ever since is one of the
six greatest in the history of complex life on earth, and the greatest
since the extinction of the dinosaurs 65 million years ago.[20]

The technological innovations during the industrial era include
farm machinery, pesticides and fertilisers. Other innovations include
all machines used in manufacturing, as well as such better-known inventions as the telephone, the electric light, the radio, television, the
automobile and the aeroplane. Many of these innovations have been
dependent on the prior invention of the internal combustion engine.

The industrial era also brought with it a further increase in the
human population, a truly tremendous one which is still continuing.
But where the population of the world has increased by a factor of

---

[18] Wilkinson (1973), p. 187.
[19] Ponting (1991), pp. 162–163; and Sessions (1995), p. xv.
[20] E. O. Wilson, as cited in Gowdy (1998), p. 66.

five since the 1850s, its consumption of energy has increased *sixty-fold* during the same period. For the wealthy 20 per cent of the world's population the industrial revolution has also meant an increase in leisure: fossil fuel resources not only constitute a tremendous surplus of energy, but one which is unequally distributed.

VCP Explanation of the Industrial Era

By the end of the agrarian period, about 250 years ago, the vicious circle had taken a gigantic turn, and was about to take one of even greater magnitude. The only way such a huge population as existed at the end of the agrarian era could continue to be supported were if a major new resource could be made available. As it turned out, such a resource existed in the form of *fossil fuels*.

On the VCP, at the time of the industrial revolution the experienced *need* for a source of energy to replace wood led to the *technological innovation* of the steam engine, among other devices, in order to make available *resources* that were previously *inaccessible*. These resources took the form of *coal*, whose non-renewability meant that its *quantity* necessarily began to *decrease*.

As is often the case in the turning of the vicious circle, the solution afforded by the technology making it possible to replace wood with coal was only partial. Though coal could be and was used in the firing of bricks, clay for the making of the bricks was also required.

At the same time, however, we must recognise that to date the most important resources made available by technological development have existed in much greater quantity than the resources they have replaced. Each major turn of the vicious circle through the horticultural, agrarian and industrial revolutions has resulted in a surplus of reserves, thereby not only supporting the then-current population, but allowing it to continue to grow. What is important here from the point of view of the VCP is that the recurrence of such a surfeit of resources has been a matter of chance, and nothing guarantees that technological innovation will make available a sufficient quantity of resources even simply to *replace* fossil fuels when they begin to dwindle or cannot be used for other reasons. On the contrary, as we continue to exhaust our non-renewable resources the probability of finding replacements is constantly diminishing; and our current knowledge suggests that no such new resources exist. The major alternative, that made available by atomic fission, presently supplies only about five per cent of the energy used in the world, and shows itself probably to require as much energy to clean up after as it provides itself – *if* it can be cleaned up after. In any case, even at this relatively low level of energy production, reserves of uranium can be expected to last only a few more decades. And so-called renewable sources of energy, among other things, will never be able to provide nearly as much power as has been provided by fossil fuels; nor, unlike oil but like nuclear energy, will they exist in handy, storable, liquid form. And neither nuclear nor 'renewable' sources of energy can power the machinery or supply the fertilisers and pesticides necessary to maintain the form of agriculture required to feed six billion or more people. This being the case, we can expect that the *next* major revolution in human development (which might be termed the *ecological*) will be associated with the diminution of fossil fuel resources, and that it will be the first such revolution that, rather than result in a *surplus*, will be connected to a general *decrease* in available resources.

As suggested by the VCP, the industrial revolution, with its requirement of greater effort to obtain the newly available resources, also brought with it a lowering of the *quality of life* of the common people. This led to *migration* on their part, first to the cities, where the new technology was being employed, and, when paid work could no longer be found there, to the periphery, i.e. the colonies, taking knowledge of the new technology with them and using it when necessary.

The compelling force of experienced *need* is noteworthy here, for it was not before they felt pressed by need that e.g. Americans began employing the sophisticated technology with which they were already familiar. As suggested by Wilkinson, this should be seen in a broader perspective as a response to ecological pressure. It is the inability of the ecosystem to maintain a particular population level that leads to our employing new methods to further exploit that same ecosystem.

We can extend this line of thinking to the whole development of humankind, and say that the changes associated with that development did not come about immediately upon the invention of new technology, but that new technology was invented and used only when there was an experience of need.

The increase in *war* during the industrial era can be explained on the VCP as the result of a combination of factors, including the constant growth of the human population. However, even though wars have become ever more devastating through the industrial period, their role as checks to population growth has (as yet) been minimal. The reason for this on the VCP is that the tremendous amount of energy available in the form of fossil fuels has, primarily through its application to agriculture, made it possible to support a *growing* population, so that the need for population control has not made itself felt to the same degree as it would have otherwise. Here the role of the powerful in society is also important, for it is they who set the trend, and to date overpopulation has been detrimental only to the weak, while benefiting the powerful.

As we move round the vicious circle thanks to our innovativeness and the existence of resources amenable to that innovativeness, the situation in which we find ourselves becomes ever more complex. This is in large part due to the magnitude of the surpluses – particularly agricultural – our technology has made available. These huge surpluses, first in the form of horticultural products and then grain in the horticultural and agrarian periods, and then in the form of food more generally in the industrial era (thanks to technological innovations in agriculture and fishing), have supported a constantly increasing population. The human population as a whole, like the populations of other species, is genetically adapted to the immediate exploitation of its available resources which, given its size and the nature of human intelligence, has resulted in specialisation and order. This complexity increases as the population grows and the specialisation required for handling the new technology increases.

When it comes to *need*, the increasing complexity and stratification of society has led to the perceived needs of the powerful, wealthy part of the population playing a role much greater than their relative numbers would otherwise suggest. In vulgar terms, it means that the greed (perceived needs) of the powerful far outweighs the basic needs of the weak.

Thus we see through the course of human existence a constantly increasing trend towards the development of technology in response to the perceived needs of the wealthy and powerful. The steam engine was not developed for the benefit of the poor, but for that of the owners of the coal mines. And the influence of the powerful has been such that the worldview in which their perceived needs may be met has come to be shared by the weak. Thus we see today, for example, a general acceptance of the need for *economic growth* – which benefits the powerful – despite the fact that common sense indicates that such growth cannot continue indefinitely (in terms of the VCP: it requires the existence of a *surplus*), and that its continued pursuit will only worsen the situation that arises when it is no longer possible.

That economic growth has been possible through the whole of the industrial era is due to the availability of huge though finite quantities of fossil fuels. This has taken the form of the constantly increasing production and construction of material artefacts, which in turn has not only meant a reduction in resources, but when applied to agriculture and housing has prompted population growth and has more generally resulted in ecologically regressive changes, including constantly increasing pollution.

In sum, during the industrial era the vicious circle has moved from the experienced *need* of a particular *resource*, namely *energy*, to *technological innovation*, making that resource available, to the *exploitation* and consequent *depletion* of the resource, which provides a *surplus* of available energy, which leads to *population growth* and *war*, and *economic growth* and *increased consumption*, and on to *increased waste*, and finally to an impending period of *need*.

The human population has increased from about 790 million in 1750 to over six billion today. Virtually the whole of this almost eightfold increase is being maintained by the use of fossil fuels to increase agricultural production and housing. When this non-renewable source of energy is no longer available in increasing quantities, either because of the environmental effects to which its use gives rise, or because it begins to become exhausted, the world will be facing a situation in which billions of people will experience real, basic need, a need which will only be reduced through a drastic reduction in the size of the human population – or eventually be eliminated through the extinction of the species.

3. CONCLUSION

The theory of human development provides a coherent view of the whole of the development of humankind. Each of the major revolutions we have undergone can be understood against the background of the VCP, as can the intervening stages involving population growth, war and energy use, as well as the phenomena of resource depletion and environmental destruction.

According to the theory, we humans are developing in a way that is clearly unhealthy as regards our survival as a species. We are headed towards a dead-end, and it may be that only by shifting to a fundamentally different worldview – to a worldview which would do well to incorporate the vicious circle principle – that we will feel sufficiently motivated to begin taking real steps in a healthier direction.

THE DEMARCATION OF MODERN SCIENCE FROM MAGIC, ASTROLOGY, CHINESE MEDICINE AND PARAPSYCHOLOGY

Given the way of conceiving of modern science presented in this volume, namely as an epistemological endeavour emanating from a core of particular ontological metaphysical principles, whether some particular such endeavour is scientific will depend on just which ontological principles it presupposes.

It seems to me that this way of approaching the question of what is or is not (modern-)scientific is much more realistic than e.g. empiricist and Popperian alternatives. For example, the view that science is based on such principles as verifiability or falsifiability (which are *methodological*) is, among other things, unable to account for the fact that many subjects whose results are verifiable (or confirmable) and/or falsifiable in the everyday sense are nevertheless not considered to be parts of modern science, and that many elements that *are* accepted as being parts of modern science, such as hypothetical theories, are neither verifiable nor falsifiable.[1]

On the view developed in the present volume, the principles from which modern science emanates are those of uniformity, substance and causality, each of which is given a physical interpretation. Thus we can look at such practices as astrology and Chinese medicine to see whether they can be considered to be sciences in the same sense as modern science by comparing their ontological presuppositions with those of modern science. In what follows of this appendix I

---

[1] To this it may be added that the empiricist notion of confirmation is inherently flawed, and that Popper's conception of falsification is incoherent for reasons having to do with an inconsistency in his philosophy of science; overlooking this inconsistency, Popper's philosophy of science nevertheless leads to scepticism: in these regards see my (2007), pp. 9, 11–14 and 38. Furthermore, even a coherent notion of falsification is not only inapplicable to theories but also to empirical laws: in the latter regard, see *this volume*, p. 82, n. 13.

shall make such a comparison, which should show not only that such practices *are* fundamentally different from science – as our intuition would suggest – but that the *extent* to which they differ can also best be understood in terms of their differing ontological presuppositions.

1. MAGIC

Belief in magic seems to be a stage virtually all human cultures have gone through or are going through. It includes such beliefs as that there is an all-pervading magical animus that manifests itself in various ways and at various places under certain conditions, some of which are controllable. Thus a person who believes in magic would believe e.g. that under particular circumstances one can influence distant events by manipulating a surrogate of that in which the influence is to be manifest (e.g. that it is possible to hurt or kill someone by sticking pins in an effigy). He or she might also believe that dealing in a certain physical way with something someone has come in contact with (under appropriate conditions), such as a piece of clothing, will have a physical effect on the person in question.

Magic, unlike modern science, is concerned with *getting things done*, and in this regard resembles *technology*;[2] or, if it is assumed actually to work, is in this regard a *form* of technology.[3] Noting this, we recognise that belief in magic nevertheless presupposes particular beliefs regarding the nature of reality. Various attempts have been made in the literature on magic to express these beliefs in terms of what have been called the *principles* or *laws* of magic,[4] which are principles in the same sense as this term is used in the present work. More particularly, they are ontological principles concerning the nature of reality, including how it operates.

According to Marcel Mauss, magic rests on *three* principles, all of which are such that causes can be transmitted along a *sympathetic chain*. These are the principles of *similarity*, *contagion* and *opposi-*

---

[2] Mauss (1950), p. 141.

[3] According to Paul Masson-Oursel, the *first expression* of technology (Ellul, 1964, p. 25).

[4] Best known to me in this regard are the efforts of E. B. Tylor (1871), Sir James Frazer (1922), and Marcel Mauss (1950).

tion: action on like produces effects on like; things that have been in contact share the same essence; and action on opposites produces effects on the other.[5] By taking these three principles together we obtain what Lawrence Jerome and others[6] call *the principle of correspondences.*

The Principle of Substance

Mauss suggests that one particular notion, the essence of which is captured in the Melanesian notion of *mana*, and which may be seen as corresponding to the animus referred to above, is of central importance to magic as a whole. Regarding it, he says:

> *Mana* is power *par excellence*, the genuine effectiveness of things which corroborates their practical actions without annihilating them. This is what causes the net to bring in a good catch, makes the house solid and keeps the canoe sailing smoothly. On farms it is fertility; in medicine it is either health or death. On an arrow it is the substance which kills . . . .
>
> [A] concept [like that of *mana*], encompassing the idea of magical power, was once found everywhere. It involves the notion of automatic efficacy. At the same time as being a material substance which can be localised, it is also spiritual.[7]

Mana is a kind of magical *potential*:

> What we call the relative position or respective value of things could also be called a difference in potential, since it is due to such differences that they are able to affect one another. It is not enough to say that the quality of *mana* is attributed to certain things because of the relative position they hold in society. We must add that the idea of *mana* is none other than the idea of these relative values and the idea of these differences in potential. Here we come face to face with the whole idea on which magic is founded, in fact with magic itself.

According to Mauss, not only is the idea of mana more general than that of the sacred, but the sacred is a species of mana. Thus, as

---

[5] Cf. ibid., p. 64.
[6] Cf. Jerome (1975), p. 43.
[7] Mauss (1950), pp. 111, 117; next quote, p. 121.

far as magical ritual is concerned, mana is the *substratum of the whole*.[8]

This conception of mana is similar to the category of substance as substratum conceived in its most universal form (so as to include, e.g., Spinoza's conception of God). Its similarity to the scientific notion of substance in particular is suggested by Mauss' referring elsewhere to the notion of mana as a category very close to those of *physis* and *dynamis* in Greek thought.[9] On Mauss' description, mana is conceived to be inherent in the world of magic in a way similar to that in which energy is today conceived to be inherent in the world of modern physics. Both notions have causal aspects, and both are intended to refer to something the total quantity of which remains constant, but whose distribution is constantly changing. A difference between them, however, is that while on the scientific metaphysics everything physical *is* energy, on the magical metaphysics mana is not *identical* with what is, but is *immanent* to it.[10] And, of course, mana is essentially spiritual, while energy in science is physical.

The Principle of Causality

In modern science the principle of causality, which states that change is caused, is refined to be the *contiguity* principle of causality.[11] Note the difference between *contagion* and contiguity. According to the contiguity principle, there is no time or space between a cause and its effect, the most common example being the everyday apprehension of the collision of billiard balls. This notion can also be applied to causal relations between a non-physical subject and a physical object, in both directions. Thus we can say, for example, that the mind (subject) is contiguous with that part of the body, e.g. the brain (object), that it is supposed to affect, and that brain states are contiguous with the mind states we may conceive of them as causally influencing.

---

[8] Ibid., p. 119.

[9] Cf. ibid., p. 117.

[10] Ibid., p. 111. This distinction is reminiscent of that between the real essences of the chemical elements as compared to those of biological species: cf. *this volume*, p. 156.

[11] Cf. ibid., pp. 59&n.–60 and 101&n.

Modern science has had great difficulty fulfilling the demands of the contiguity principle. As a consequence, a weaker principle, that of *proximity*,[12] has generally been accepted. According to the proximity principle, the nearer a cause is in space (or time, though not paradigmatically) to its effect, the greater the effect will be. An example is that of Newtonian gravitation, in which the cause (attraction) stretches out through empty space, and is stronger (according to the inverse-square law) the nearer the object is upon which it has an effect. All the same, proximity has never been accepted as completely satisfactory in science, as is evident e.g. from the fact that physicists are still attempting to find a means by which gravity operates by contiguity.[13]

Thus we can see magic as being similar to modern science in presupposing a principle of causality, but differing with regard to the form that causality takes according to its principle. According to the paradigmatic modern-scientific notion, causes are not only contiguous to their effects, but are also *efficient* and *physical.*

The scientific principles of contiguity and proximity may be contrasted with both the similarity principle and the contagion principle of magic. Given similarity, the requirement of contiguity need not be met. The place where the effigy exists when it is pricked may be far distant from the place of the pricking's anticipated effect.

Similarly, the contagion principle of magic also allows action at a distance, though it at the same time requires contiguity, but of a weakened sort. The contagion principle, in distinction from the contiguity principle, allows that the occurrence of the spatial contiguity *precede* that of the cause-effect action. The physical entity contiguous with the intended object was so at one time, but not at the time when the magical cause – e.g. the burning of the entity – is to give rise to its effect.

---

[12] Cf. ibid., p. 60. It may also be noted that the acceptance of non-locality in quantum mechanics would suggest that there, at least, not even the proximity principle is met, and we would have an instance of pure action at a distance in the core discipline of modern physics.

[13] In this regard, cf. ibid., pp. 101–102. As expressed by D. R. Griffin, the desire still exists today "to find explanations of gravitation that do not involve attraction at a distance, such as curvature of space and 'gravitons.'" (1996, p. 94).

The human sciences accept the modern-scientific notion of causality but admit another notion as well, namely that of causes having an *intentional* or *teleological* (rather than *efficient*) ground in human *agents*.[14] Magic accepts these two notions and admits a *third*, which is that of acts performed by agents which are *not* contiguous with their effects. In other words, as already noted, magic accepts *action at a distance* by a *subject* (agent).[15] In fact it may be said that this notion of causality is *paradigmatic* for magic. Thus we see that the principle of causality for magic differs from that for modern science both in its conceiving of causally active *subjects*, as do the human sciences, and in its accepting pure action at a distance, which the human sciences do not. And on top of this we might mention that while interest in causality for modern science is purely *epistemological*, for magic it is *utilitarian* (magic as a form of technology).

Distinct Applications of Intention

As regards the linking of the situation in which the cause takes place to that in which the effect is to take place, E. B. Tylor has suggested that what is important is that *some* association can be seen to exist between the two, i.e. that the principle of correspondences can be applied. As expressed by Mauss, the seeing (making) of this association is "accompanied by [among other things] direction of intent."[16] Thus the practitioner *intends* that there be an association or correspondence between the surrogate of an object and the object itself. Here the notion of intention can be meaningfully applied only

---

[14] Cf. *this volume*, p. 140.

[15] As regards this notion of action at a distance, it should be pointed out that even in magic various means of operation have been envisaged which may avoid it. For example there is the idea of *effluvia* which leave the body, magical images which travel about, and lines linking the magician and what he acts upon; even his soul can leave his body to perform magical acts (Mauss 1950, pp. 72–73). It might also be possible to conceive of magical effects' being transmitted through some ubiquitous or immanent magical stuff such as mana. In this way the performance of an act intended to have an effect on something or someone far distant might be seen as a matter of touching on the magical essence of reality, and directing it in a certain way.

[16] Ibid., p. 68.

if we assume that the relevant correspondence does not in fact exist. This application of intention I shall call the *first kind* of application.

The *second kind* has already been mentioned above; it is manifest when the magician *acts* so as to create a particular effect. Thus his or her *intention* is that by acting on the surrogate in a particular way, an effect will be brought about in the object.

Note the position taken in the Western worldview of today with regard to this second kind of application, namely that there do exist intentional or teleological causes as are treated in the human sciences (otherwise, for example, ethics and the idea of responsibility would make no sense), while the spiritual action at a distance believed to be possible in magic is considered impossible – despite our admission of *physical* action at a distance, albeit in keeping with the proximity principle.

The Principle of the Uniformity of Nature

The principle of the uniformity of nature says that natural change is lawful, and it can take more or less strict forms, the strictest of which is determinism.

In *The Golden Bough* James Frazer suggests that magic is perceived by its practitioners as operating according to a deterministic form of uniformity, such that operations on a surrogate *must* have an effect on (or a tendency to affect) the object:

Wherever sympathetic magic occurs in its pure unadulterated form, it assumes that in nature one event follows another necessarily and invariably without the intervention of any spiritual or personal agency. Thus its fundamental conception is identical with that of modern science; underlying the whole system is a faith, implicit but real and firm, in the order and uniformity of nature. The magician does not doubt that the same causes will always produce the same effects, that the performance of the proper ceremony, accompanied by the appropriate spell, will inevitably be attended by the desired result, unless, indeed, his incantations should chance to be thwarted and foiled by the more potent charms of another sorcerer.... Thus the analogy between the magical and the scientific conceptions of the world is close. In both of them the succession of events is assumed to be perfectly regular and certain, being determined by immutable laws, the operation of which can be foreseen and calculated precisely; the elements of caprice, of chance, and of accident are banished from the course of nature.[17]

Thus, assuming that a deterministic form of the uniformity principle is correct, Frazer says: "the fatal flaw of magic lies not in its general assumption of a sequence of events determined by law, but in its total misconception of the nature of the particular laws which govern that sequence."

Frazer's suggestion of deterministic uniformity in magic, which was not adopted by his predecessor Tylor, seems however to have been rejected in the literature,[18] it now being thought that magic presupposes a weaker version of uniformity, of the sort as is normally presupposed in conceiving of intentional action (e.g. by Aristotle and in the human sciences), and as is not to be found in the core disciplines of modern science.

While the form of uniformity presupposed in magic may be weaker than that of the deterministic uniformity principle paradigmatic for modern science, the key differences of principle between the two endeavours concern the principles of *substance* and *causality*, where integral to magic is the idea of *spiritual* substances capable of *acting at a distance*.

2. ASTROLOGY

According to such authors as Mauss[19] and Jerome, astrology is a form of magic. This view is supported by looking at the two enter-prises in terms of their core principles. According to Jerome, what makes astrology in its modern, i.e. Hellenistic, form a system of magic is essentially its adoption of the principle of corresponden-ces.[20]

Emending Jerome's reasoning somewhat, I suggest that we can see the operation of the principle of correspondences in astrology if we compare the relation between a *surrogate* and the intended *object*

---

[17] Frazer (1922), pp. 48–49; next quote, p. 49.
[18] Cf. e.g. Tambiah (1990), pp. 52–53.
[19] Mauss (1950), p. 46.
[20] Jerome (1975), p. 37.

of a magical operation with the relation between a particular constellation or planet and particular human beings. In each of these cases the relation involves a particular correspondence which either *paves the way for* (magic) or *constitutes* (astrology) a causal link between the two. The situation is somewhat complex, since we have to do both with the relation between e.g. a constellation and a particular person, and with the astrologically relevant qualities they each have. I am inclined to say that the correspondence might best be thought of as existing between the constellation and the person *thanks to* their respective qualities, though I expect an argument could be made for the correspondence being thought to exist between the qualities rather than their bearers. In any case, assuming that there in fact exists no real such correspondence, it is the *intention* (first kind of application) of the believer in magic or astrology that puts it there (as an 'intentional object').

The two cases differ however with respect to the second kind of application of intention. Where the magician intends to bring about particular effects e.g. through the power of his or her will, in the case of astrology this use of intention is absent. This difference is expressed by Jerome as follows: "In most sympathetic magic, the magician's strength of will is supposed to complete the magical link between amulet and corresponding object; only in astrology is the magical link made automatically through the 'celestial harmonies of the spheres.'"[21]

As regards the first kind of application, on the other hand, the parallel is clear. The correspondence assumed to exist between an effigy and an object in magic is of essentially the same sort as between a constellation and a person (or persons) in astrology. And to this we can add that an integral part of astrology, like magic, is the employment of the notion of action at a distance.

But, as is manifest in their different applications of intention, where magic might be considered a form of technology (or an attempt at a form of technology), astrology is so only indirectly. The magician, through particular manipulations, believes himself able to bring about particular effects. The astrologer, through certain divina-

---

[21] Ibid., p. 43.

tions, can only indicate predetermined causal relations. The astrologer cannot manipulate the influences himself, or make them act or cease to act. Nevertheless, these influences are not immutable: being aware of them can in certain cases allow one to avoid or deflect them.

I should say that the core principles of the astrological worldview or perspective are radically different from those of modern science. The modern conception of the earth's non-special place in the universe is quite in keeping with the metaphysics of ancient atomism,[22] which in turn is the fundamental metaphysics of modern science. Astrology, on the other hand, is based on a metaphysics that is more similar to Aristotle's. Like Aristotle, astrology both takes the earth to be at the centre of the universe and admits causes that transcend the purely physical – in the case of human action, more particularly *teleological* causes.

Thus if we look at the basic principles of modern science and compare them with what might be considered to be the principles of astrology we should see clear differences. Let us begin with the *uniformity* principle. In modern science this principle applies only to physical entities, so in the case of astrology we must admit that this requirement is not met, since astrology also has to do with non-physical correspondences and self-willed spiritual entities (human agents). But as regards uniformity itself, the difference is not so great. Just as in the paradigmatic deterministic view of modern science, on the astrological conception the influence of the heavens on people's characters is constant. And, where the spiritual influences of the heavenly bodies can be diverted, so can deterministically acting causes in modern science – though in science this in itself requires the concept of a spirit (of e.g. an experimenter) to so divert them.

As regards *substance*, there appears not to be any substratum for astrology, unless one takes the realm in which the spiritual forces of the heavens exist (perhaps what Jerome refers to as the 'celestial harmony of the spheres') to be a substratum. In any case, unlike modern science, astrology admits spiritual substances.

---

[22] Cf. Lindsay (1971, p. 360): In Roman times "[o]nly the Epicureans made a consistent effort to build a world-picture from which astrology was excluded."

As regards *causality*, whether or not one takes astrology to be a form of magic, we have a situation similar to that in the case of magic, where causes can emanate from spiritual entities, as well as act at a distance.

3. CHINESE MEDICINE

Chinese medicine, like any medicine, is distinct from modern science and other purely epistemological activities in that its primary aim is not to gain knowledge or understanding but to *heal* or *prevent illness*. Like magic, its purpose is thus utilitarian, and, to the extent that it works, it may also be seen to be a form of technology. Nevertheless Chinese medicine, like both magic and modern Western medicine, presupposes certain principles regarding the nature of reality and how it operates, and can on this basis be compared with modern science. In this regard we might first note that the categories most central to the fundamental presuppositions of Chinese medicine are those of *qi*, and *yin* and *yang*.

Qi, variously translated as 'vital energy' or 'vital substance' (note the implied animism), is central to all aspects of traditional Chinese medicine; and the same can be said of the distinction between yin and yang.[23]

As regards yin and yang:

The basic concept . . . is that everything in the universe follows an alternating cycle, e.g., day into night, growth into decay. Although each phenomenon is opposite, each also contains an aspect of the other . . . . In the body there is a continual balance between yin and yang.[24] Yin and yang are interdependent. Extreme yin will change to yang and vice versa.[25]

---

[23] Norris (2001), pp. 1, 6.

[24] The idea of health as consisting of a state of balance was the common view in ancient Greece, manifest e.g. in the teachings of Hippocrates and Aristotle, and prevailed in Europe throughout the Middle Ages. The clearest Western correspondent to the yin/yang distinction itself is perhaps Empedocles' opposition of Love and Strife.

[25] Ibid., p. 6. Cf. also Chang (1976, p. 67): The *Nei Ching* states that "[t]he entire universe is an oscillation of the forces Yin and Yang." Here we see that these categories underlie a whole worldview, not just Chinese medicine – just as the central

Yin and yang are thus conceived to be the two main determinants of various types or aspects of qi. Among central such aspects are, in the case of yin, *the solid*, which is produced by the internal organs to nourish the body; and in the case of yang, the *energy* to move the solid, which is the functional activity of the organs themselves.[26] From this point the situation becomes successively more complicated, but all the time with the categories of qi and yin and yang being manifest in various ways.

In Chinese medicine, an imbalance in the amounts of yin and yang is taken to be the cause of all pathology. Thus an ability on the part of practitioners to recognise the aspects of a disease as pertaining in particular ways to yin and/or yang is to help them understand its nature and be better equipped to administer a treatment that leads to recovery.[27]

It is not difficult to see a clear correspondence here with the fundamental principles of modern science. Qi, like the mana of magical thinking,[28] corresponds to the modern-scientific category of *substance*, while yin and yang together correspond to *cause*. In the case of Aristotelian science, yin might correspond to the material cause, while yang corresponds to the formal cause. As regards modern science we might perhaps say that yin corresponds to an attractive force, while yang corresponds to a repulsive force. And as regards *uniformity*, it is clear that a form of lawfulness is assumed in Chinese medicine.

While these similarities exist, there are also of course important differences between the categories and principles of Chinese medicine and those of modern science. Central to the differences is that the latter concern only *physical* reality.

---

categories of modern Western medicine are essentially the same as those of the more general materialistic-scientific worldview of our times.

[26] Norris (2001), p. 2.
[27] Chang (1976), pp. 69, 72.
[28] And like the Hindu notion of *bráhman*: cf. Mauss (1950), pp. 116–117.

4. PARAPSYCHOLOGY

The two central fields of parapsychology concern what is termed *psi-gamma* (the acquisition of knowledge or information by non-sensory means or without physical mediation) and *psi-kappa* (the affecting of physical objects by a subject without physical mediation). Psi-gamma phenomena include *telepathy* (the sending of thoughts from one mind and the receiving of them by another without physical mediation), *clairvoyance* (knowing independently of sensory input that something is the case), and *precognition* (knowing on non-sensory grounds that something will occur). Psi-*kappa* phenomena are co-extensive with those of *telekinesis*, and include the *psychic healing* of physical ailments. Other aspects of parapsychology concern *out of the body experiences* and *the survival of bodily death*, the first of which may involve psi-gamma, and the second psi-gamma and/or psi-kappa.

The Principle of the Uniformity of Nature

In modern science the uniformity principle is manifest on the empirical level in the *repeatability of experimental results.* That the results of experiments in parapsychology have not been consistently repeatable[29] has consequently been seen as an indication that parapsychological studies ought not be accepted into the realm of science. Note the ontological nature of this line of thinking. It suggests that *despite* parapsychology's scientific methodology, it ought not be admitted to science since its *subject-matter* has not been conclusively shown to follow the uniformity principle.

Though parapsychology, like the human sciences, does accept (a form of) the uniformity principle, the nature of its experiments, involving constantly changing human and animal subjects and an experimenter who is part of the experiment, is such that it is virtually impossible to provide the preconditions for the repeated manifestation of causes in a uniform way. The same may be said of the human

---

[29] One of the most successful highly sanctioned runs of empirical tests suggesting the existence of psi-phenomena is J. B. Rhine's, where his best subject averaged eight hits per run (chance = five) over a total of 17,250 guesses. (Hyman 1985, p. 45).

sciences however, so in this particular respect parapsychology need be no further from being a modern science than they are. The reason for parapsychology's not having the status even of the human sciences must lie elsewhere. As expressed by H. L. Edge: "Parapsychology fails to be a science for the same reason that voodoology would not be considered a science, and I think it is not because of the lack of replication or prediction."[30] And, as noted above, it is not due to a difference in methodology. As we shall see, as regards its core principles parapsychology is not so different from "voodoology," the fundamental 'failure' to be sciences in both cases being ontological and depending primarily on the principles of substance and causality.[31]

The Principle of Substance

When it comes to the category of substance as substratum, if parapsychology is taken to have such a substance it is clear that it should be more similar to the spiritualistic or animistic substances of magic and Chinese medicine than to the physicalistic substance of modern science. It turns out that parapsychology does have a correlate for

---

[30] Edge (1985), p. 61.

[31] Since in this appendix our interest is in demarcating science from non-science, in what follows we shall concentrate on the differences and similarities of parapsychology and modern science. But an important question related to this concerns the viability of parapsychology *per se* with respect to its very *intelligibility.* (Regarding the question of intelligibility in the case of modern science, see *this volume*, Section 5 of Chapter 4.) Thus we might say that where the basic principles of modern science afford an intelligible conception of change, it may not be possible to say the same of the principles of parapsychology. In this general regard Edge, for example, says: "The failure of modern parapsychology is not that we fail to have replication, nor is it that parapsychology studies non-existent phenomena; rather, it is that we have not made our phenomena intelligible. That is why parapsychology is not a full-blown science. Repeatability is only a problem insofar as it has become a symbol for this failure." (1985, p. 64).

At the same time, however, intelligibility presupposes the existence of a particular worldview in the context of which what is being considered is to be deemed intelligible or otherwise; and while it is possible that parapsychological phenomena may not be intelligible against the background of the modern Western worldview, this need hardly be the case if it is considered against the background of e.g. the worldview of magic.

mana, qi, etc., and it is *psi*, of which psi-gamma and psi-kappa are the two fundamental forms. In this regard then we can fairly align psi with mana and qi.

As regards parapsychology and the notion of substance, however, much greater interest has been shown in the question of whether the *self* or *soul* is a substance, such as to imply e.g. that it could survive bodily death. In this case substances are conceived of as being discrete (Platonic) entities rather than as constituting an all-pervasive (spiritual) substratum. A connection might be made between the two conceptions, however, if we were to regard our selves or souls as *participating* in a spiritualistic psi, and each of us as being one of its manifestations.

The Principle of Causality

That we are correct in looking for the criteria for demarcating modern science from other epistemological endeavours on the basis of their respective *principles* finds support not only in others' analyses of magic, but also in C. D. Broad's analysis of parapsychology. In it he compiles a list of what we may consider as candidates for being real fundamental causal principles of the worldview generally accepted today, which he terms "basic limiting principles." These principles are similar to what Thomas Reid presents as the principles of common sense.[32] Antony Flew in fact considers Broad's principles to "have been accepted as items of basic common sense," and furthermore to be "prior to and more fundamental than any named laws of physics."[33]

Broad's principles include that:

1.1. It is impossible that an event should begin to have any effects before it has happened.

1.3. It is impossible that an event should produce an effect at a remote place unless a finite period elapses between the two events, and unless that

---

[32] Cf. Reid (1764).

[33] Flew (1987a), p. 37. In this regard, cf. George Price: "The conflict [of ESP with current scientific theory] is at so fundamental a level as to be not so much with named 'laws' but rather with basic principles." (1955, p. 217).

period is occupied by a causal chain of events occurring successively at a series of points forming a continuous path between the two places.

2. It is impossible for an event in a person's mind to produce *directly* any change in the material world except certain changes in his own brain.

3. A necessary, even if not a sufficient, immediate condition of any mental event is an event in the brain of a living body.[34]

Principle 1.1 is more a principle of the modern Western worldview than of modern science,[35] while 1.3 has a close affinity to the scientific principle of contiguity. Principles 2 and 3, on the other hand, with their references to minds and mental events, involve more than is captured by the physicalistic principles of modern science.

Whether or not the soul is taken to be a manifestation of psi, souls or selves in parapsychology are seen to constitute *causes*, as do physical entities – manifestations of substance – in modern science. An important difference however is that in parapsychology, as in magic and the human sciences, the transmission and/or reception of signals are conceived as being the effects of the operation of the *spirit*, not the *body*.

Before proceeding further we might first consider four types of causality or *four categories of cause* (not intended to be either exclusive or exhaustive), all of which we have dealt with earlier, as regards the extent to which they may be considered to be essentially modern-scientific, applicable to the human sciences, and/or applicable to parapsychology. I shall list them here with short descriptions or examples, some of which have already been mentioned above:

A. *Contiguous, efficient, deterministic* and *physical*, such as in the case of colliding billiard balls and Greek atoms.

---

[34] Cf. Broad (1949), p. 40; the numbering is Broad's.

[35] The scientific principle of causality does not indicate a temporal direction for cause-effect relations, and in the principle of contiguity the notion of temporality collapses; on the other hand, however, paradigmatic instances of causal mechanisms as conceived in science involve causal chains in which the initial causes precede the final effects. For an exploration of the idea that effects can precede their causes and still be in keeping with the dictates of modern physics, see Faye (1989).

B. *Non-contiguous, proximate* and *efficient*, such as in the case of gravity (so far as we know).

C. *Contiguous, spiritual* and *teleological*, e.g. human action as studied in the social sciences.

D. *Non-contiguous, distant* and *spiritual*, as in magic and parapsychology.

A. *Contiguous efficient deterministic physical causes (and effects).* The paradigm of contiguous, efficient causes on the empirical level, as mentioned in the section on magic, is the collision of billiard balls (cf. also Chh. 4 and 5); and the workings of clocks have also often been referred to in this regard. Such interactions are considered to be *mechanistic* in that they involve the *contiguity* of parts of *substance* in their *uniform (deterministic), efficient*, causal interaction. When such is the case, we may speak of the presence of a *causal mechanism*.[36] Thus we might say that the reason Newton's theory of gravitation does not indicate such a mechanism for gravity is that, despite its depicting uniform change resulting from an efficient cause, it does not represent contiguous action.[37] And, on the present view, Newton's three laws of motion are together taken to constitute his *mechanics* only if they are understood in terms of contiguous action. Here we can also state that, in essence, an *occult quality* is nothing other than a non-contiguous cause; and we can thus understand the criticism of Newton's theory for its having to do with such qualities.

---

[36] Cf. *this volume*, p. 101: "In the case where the theory in question succeeds in depicting a regular cause operating via the substance in a way which is in keeping with the contiguity principle, that part of the substance which mediates the causal relation may be termed a *causal mechanism*." In this regard cf. also Griffin (1996, p. 92): "The new metaphysics for science introduced in the seventeenth century was called, not coincidentally, the 'mechanical philosophy.' [W]e may assume that the chief point at issue in speaking of 'mechanism' was an exclusive focus on efficient causes, in distinction from 'final causation.' The real bite of mechanism...is that, by excluding all self-determination, it entails complete determinism. This was indeed one of the central issues, but not the only one. An at least equally crucial meaning of the 'mechanical philosophy' was that action at a distance was proscribed."

[37] *This volume*, pp. 101–102; cf. also p. 205.

Thus the notion of causal mechanism should also cover similar cases in which the interacting objects are not physical (in keeping with Broad's second principle). And, following the discussion of contiguous spiritual-physical causal relations in the section on magic, it would be correct to speak of causal mechanisms operating in the mind, or between the mind and body, so long as the contiguity and deterministic uniformity principles are met. In this way, the operation of yin and yang of Chinese medicine could meet this requirement.

B. *Non-contiguous, proximate causes.* In the section on magic Newtonian gravitation functioned as an example for the application of the *proximity* principle. As was said there, though on Newton's theory the requirement of contiguity is not met, the action of gravity is nevertheless *proportional to distance.* Note that the proximity principle, like that of contiguity, should also be applicable to similar cases where the causes are spiritual, or are a combination of spiritual and physical.

C. *Contiguous, spiritual, teleological causes.* While the human sciences allow spiritual actions to have physical effects, and vice versa, as also mentioned in the section on magic they do not allow action at a distance. As discussed there, and partly covered in Broad's second principle, spiritual causes are to be contiguous with their physical effects in the brain (or body more generally), and physical causes in the brain are to be contiguous with their spiritual effects.

A sub-species[38] of contiguous, spiritual causes consists of those which are *teleological*, i.e. which involve *willing* on the part of an *agent.* The operation of such causes, since it is not deterministic, is such that strictly speaking the notion of a mechanism is not applicable.[39]

---

[38] As noted by Donald Evans: "[W]hen changes in consciousness cause changes in the body this is not always an instance of agent causality." (1996, p. 57).

[39] Though it is nevertheless applied when the situation in question is conceived of deterministically, which in itself creates a tension. In this regard, cf. *this volume,* pp. 141&n. and 135, n. 9. The situation is further complicated by the fact that the

On the view presented in this book, the difference between the human sciences and those sciences whose principles lie closer to the core of modern science lies essentially in the human sciences' adoption of a notion of spiritual entities and of causality as stemming from and affecting such entities.[40] I would suggest that it is the fundamental difference of such spiritual entities or agents and physical entities, both in how they are and how they act, that constitutes the fundamental difference between the human sciences and modern science, an ontological difference which no methodology can remove. In terms of principles, it is the basic difference in the principles of *substance* and *causality* that divides the two. Furthermore, the non-deterministic aspect of teleological causality[41] further removes it from the core causal principle of modern science, a problem similar to that experienced by quantum mechanics with its indeterminacy principle.

A question that then arises, however, concerns the acceptance of the uniformity principle on the part of the human sciences and parapsychology. It may appear that the introduction of teleological causes means the forsaking of uniformity. What we find, however, is that the uniformity requirement is met on the empirical level in the consideration of *large numbers* of what are conceived to be teleological cause-effect relationships, which are such that uniformity becomes to some extent evident.

D. *Non-contiguous, distant, spiritual causes.* Thus parapsychology, like the human sciences, differs from modern science in its admitting agent causality; but it is even further removed from science in its not requiring contiguity (unlike the human sciences) nor proximity (unlike physics). More particularly, as regards psi-*gamma*,

---

effect of a teleological cause is normally at a distance from the spirit willing that effect, and thus involves a causal chain.

[40] Cf. Meynell (1996), p. 23: "In general, we have two kinds of explanation of phenomena: the physical kind best exemplified by natural science, and that involving appeal to conscious agents."

[41] This aspect consists in the fact that the striving to attain a particular end does not ensure that that end is actually reached. That an act is *freely willed* does not in itself contravene the uniformity principle *per se*, but suggests rather that there can exist (spiritual) causes that are not the effects of other causes.

the receipt of telepathic or clairvoyant information should be independent of where or when it is sent; and the physical distance between the cause and its effect should be irrelevant.

As regards psi-*kappa*, the question of *temporal* contiguity hardly arises. But as regards *spatial* contiguity, the spiritual cause and its physical effect are normally not spatially contiguous, as when someone is considered to affect the rolling of dice with their minds; and in such cases neither is the proximity principle obeyed. In other cases however, such as those involving psychic healing where light tactile pressure is employed, there appears to be contiguity, while it is nevertheless the spirit that is considered to be doing the healing.

5. METAPHYSICS AND WORLDVIEWS

As suggested by James McClenon,[42] and as is the basic presupposition of this book, the methodology of science is based on various *metaphysical* assumptions. That the present question has to do with metaphysics is also expressed by Douglas Stokes, who suggests that *materialism* is a metaphysical doctrine.[43]

The seeing of worldviews as each being based on a particular metaphysics[44] can help us understand not only the considerations of G. H. von Wright cited in the body of this book,[45] but also those of David Ray Griffin where, quoting Jerome Ravitz, he agrees that the "Scientific Revolution was primarily and essentially about metaphysics; and the various technical studies were largely conceived

---

[42] McClenon (1984), p. 3.

[43] Stokes (1985), p. 380. Cf. also Mary Hesse, who says, "there is a sense . . . in which basic categories . . . in science are always *a priori*; and the new principles [of the Scientific Revolution] were understood partly as the replacement of the old metaphysics by new, and were argued on metaphysical grounds. [T]he seventeenth century world-model according to which all physical change is *really* produced by matter in motion . . . is in a sense a metaphysic; derived from the Greek metaphysical theory of atomism; established by the overthrow of opposing metaphysical systems, namely the Aristotelian, Stoic, and Neo-Platonic; and justified by the Pythagoreanism of Copernicus, Kepler, and Galileo, and by the metaphysical arguments about primary qualities of Descartes, Galileo, and Locke." (1961, pp. 98–99, 125).

[44] Cf. *this volume*, e.g. pp. 158n., 200, 207–208, and above, pp. 250, 255.

[45] Ibid., p. 61, n. 19; quote following, Griffin (1996), p. 92.

and received as corroborating statements of a challenging world-view." And we can also understand McClenon's seeing a coupling between the a priori of metaphysics and the notion of a worldview, together with his suggesting that parapsychology opposes the scientistic worldview of our times, rendering the paranormal as a priori impossible.[46]

6. HISTORICAL DEVELOPMENT OF THE NON-PHYSICAL

According to the development of the history of ideas as presented in Chapter 10 of this volume, "three streams of Greek philosophical thought have shaped virtually the whole of Western intellectual development – philosophical and modern-scientific – right up to today." These streams are the Presocratic atomist – from which modern science has developed – the Platonic, and the Aristotelian. Regarding the last two, Broad and Richard Robinson point out that:

According to the Platonic theory, a man is primarily something immortal and imperceptible and spiritual, which for one or more short periods is united with something mortal and perceptible and material, namely, a specimen of that animal labelled 'homo sapiens' by the biologists. According to the Aristotelian theory, man *is* that animal labelled 'homo sapiens' by the biologists; and that animal is not linked to any immortal imperceptible twin (you see I am disregarding the famous little chapter in which Aristotle reverts to Platonism); and what we refer to as its soul or mind is the entelechy or form or higher behaviour of that animal.[47]

The first major step in the seventeenth-century revolution against Platonic and more particularly Aristotelian thought was the acceptance of Copernican astronomy. With the advent of Copernicanism and the demise of the Aristotelian earth-centred conception of the universe, as has been pointed out by J. V. Field, acceptance of the teachings of astrology also began to decrease markedly.

The rapid decline in the intellectual standing of astrology in the course of the 17th century is roughly contemporary with a rise in the respectability of Copernicanism. . . . If all the planets, including the Earth, were believed to be

---

[46] McClenon (1984), pp. 131, 68; quote following, *this volume*, p. 199.
[47] Robinson & Broad (1950), p. 272; next quote, Field (1987), p. 144.

moving round the Sun, the fact that a particular planet was 'in' a particular constellation merely told one something about its position relative to the Earth rather than, as in a geocentric cosmology, its actual position in regard to the Universe as a whole. While this does not refute the ascription of particular 'houses' to each planet, it does somewhat decrease their cosmic significance. Moreover, a similar weakening will be found in all the reasoning which depends upon Zodiac signs: in a heliocentric system we no longer have absolute properties of the macrocosm exerting their influence upon Man.

Once the earth was removed from the centre of the universe the way lay open for the introduction of the atomistic-physicalistic view of the Scientific Revolution. With this revolution we have a shift away from the notion of spirit,[48] one even further than that made in the move to Aristotle's philosophy from Plato's in the Middle Ages. Thus, for example, the Scientific Revolution not only eliminated Aristotelian science, but drastically weakened the position of the Church. As suggested by H. Butterfield, Newton's first law and the modern theory of motion in the seventeenth century helped to drive the spirits out of the world and open the way to a universe that runs like a piece of clockwork.[49] As expressed by Stokes, at this time the universe became a big machine governed by mechanical principles.[50]

That the Scientific Revolution reinforced the principle of contiguity in particular is further supported e.g. by Newton's continual attempts to find a physical link for gravity.[51] And it also gains support from present-day writings such as those of Mary Hesse, who suggests that the idea of action at a distance lost favour through the introduction of the mechanical philosophy of nature, according to which physical particles were purely material, having no sympathies or antipathies allowing them to exert or receive influence at or from a distance.[52] Further support comes from Richard Westfall, who claims that "All agreed that the program of natural philosophy lay in demonstrating that the phenomena of nature are produced by the

---

[48] The Scientific Revolution's elimination of the idea of what is *animate* as underlying change is emphasised by Hesse (1961, pp. 101, 111–112).
[49] Cited in Stokes (1985), p. 382.
[50] Ibid., p. 383.
[51] Cf. *this volume*, pp. 59 and 102, n. 8.
[52] Hesse (1961), pp. 118, 125 and 291.

mutual interplay of material particles which act on each other by direct contact alone."[53]

As suggested by Griffin, the development of a fully materialistic position by society at large

occurred in the latter half of the eighteenth century in France and in the latter half of the nineteenth century in the English-speaking world (thanks to a large extent to Darwin). With this development, the 'mind' was fully within nature, being purely a function of the brain (as the notorious Hobbes had suggested), and was therefore subject to the same prohibition against action at a distance as the rest of nature.

There was a backlash to this development, however, manifest in Hermetic and other 'magical' philosophies, which allowed influence at a distance as a purely natural occurrence, including that to and from minds.[54] There was also an upsurge in interest in the two traditional cult sciences of astrology and alchemy, as well as in a wide variety of older and newer systems of magic.[55] And in the academic world, where the metaphysics of atomism had developed into modern science, Platonism and Aristotelianism continued on alongside, and were modified in various ways so as to be in keeping with what science had established while retaining the notion of spirit. Leibniz' philosophy can function as an example. In it his monads are *like* atoms, only they have the teleological properties of Aristotelian philosophy; and his great hope is to invent a universal *language*, to which a formal logic *à la* Aristotle should apply.

Part of this reaction, particularly with regard to the influence of the Scientific Revolution and Darwin's theory of natural selection on the Church, was the rise of spiritualism in the second half of the 1800s – its predecessor being the Shaking Quakers. Mediums had arrived on the scene already around 1770; and with the seances of the infamous Fox sisters beginning in 1848, a spiritualist mania was born.[56]

---

[53] Cited in Griffin (1996), p. 92; next quote, p. 95.
[54] Ibid.
[55] John Beloff, cited in Alcock (1985), p. 546.
[56] Alcock (1985), p. 548; quote following, Flew (1987b), p. 14.

Essentially this same reaction was manifest at the end of the nine-teenth century with the formation of the Society for Psychical Research, the intention of its members being to support a Platonic-Cartesian point of view, which maintains

that somehow inside and controlling the creature of flesh and blood is a mind or soul or self; and that this is a substance, in the sense that it could significantly be said to survive separately – unlike, say, a personality or a temper or a grin; that this substance is incorporeal, immaterial, and somehow non-physical; and, finally, that it is the essence or core or actual person.

As regards philosophical development, after the Scientific Revolution and up until the twentieth century it consisted in attempts to resuscitate the notion of the spirit in a form in keeping with either Plato or Aristotle while not obviously conflicting with the results of science. This line has continued in the twentieth century and up to the present in the Continental tradition, while independently of it the analytic tradition has developed, consisting in a misconceived Aristotelian-logicist kowtowing to science.

In general the reaction to scientific materialism has been weak, however. The success of modern science has not only meant a general decline in interest in the spiritual, but the acceptance of the principles of science as sacrosanct, thereby discouraging the investigation of fimdamentals, and accounting for the peripheral place of philosophy in our universities.

Given the scientistic spirit which has enveloped us since the time of the Scientific Revolution, we can see the strong emphasis on measurement in the human sciences during the twentieth century as being a manifestation of psychologists' and social scientists' attempts to have their subjects be accepted as parts of modern science, where emphasis is placed on the manifestation of the uniformity principle on the empirical level due to the influence of positivism in the form of behaviourism.[57] This phenomenon is also evident amongst parapsychologists during this same period, in their attempts to create replicable experiments.

---

[57] In this general regard, see *this volume*, pp. 130–131.

7. MODERN SCIENCE AND THE SPIRIT

The fundamental problem for modern science with regard to the spirit is evident already in early Greek atomism, with its lacking categories for the self and psychic states.[58] This problem remains in modern science, both as a paradox with respect to the nature of its own activities, as well as a major lacuna with respect to what it is capable of explaining. Thus modern science presupposes an agent for its own existence;[59] and at the same time the spiritual element generally acknowledged to exist in human activities cries for explanation. Science, limited as it is to physicalistic categories, cannot handle either of these issues.[60]

As expressed by J. B. Rhine with respect to the latter point, and as is in keeping with what I parenthetically suggested regarding ethics and the idea of responsibility in the section on magic:

Under a mechanistic determinism the cherished voluntarism of the individual would be nothing but idle fancy. Without the exercising of some freedom from physical law, the concepts of character, responsibility, moral judgment, and democracy would not survive critical analysis. The concept of a spiritual order, either in the individual or beyond him, would have no logical place whatever. In fact, little of the entire value system under which human society has developed would survive the establishment of a thoroughgoing philosophy of physicalism.[61]

Hugo Meynell, for his part, asks:

How, if at all, does one adapt 'scientific' inquiry, for example, to the treatment of ethical, metaphysical or religious questions? Are the methods of the natural sciences to be extended without modification to what are sometimes called the human sciences? If they are not, what types of modification are needed, and why? Questions like these can hardly be avoided if one is going to consider applying 'scientific' methods to the matters in which we are interested here.

---

[58] Cf. ibid., pp. 201–203.
[59] As suggested by Evans (1996, pp. 55–56, 58–59).
[60] As pointed out regarding the latter on p. 91 of *this volume*; cf. also p. 60: the core principles of modern science "can be criticised for difficulties had in applying them to psychological phenomena, conceived quite generally."
[61] Rhine (1954), p. 32; next quote, Meynell (1996), p. 24.

In this regard Stokes suggests that present scientific theory is incomplete when it comes to explaining mental phenomena: "There is no understanding at present of how or why certain patterns of neural activity give rise to conscious experience."[62] I should say that this is not a problem just "at present," but is a problem *in principle*, and lies in the nature of the metaphysical core of science.[63] Elsewhere Stokes suggests that strict materialism contradicts the fact that one has direct experience of such mental events as sensations, thoughts, memories etc., and that psi phenomena threaten the worldview of scientists, which is clung to with almost religious tenacity.[64]

In the same vein, James E. Alcock rhetorically asks how a science of the spirit can exist, given that science by its very nature is materialistic,[65] while B. Mackenzie and S. L. Mackenzie claim that anti-materialism is part of the identity of the paranormal, and that if materialism is incorrect, once this is realised the implications for humankind could be overwhelming.[66]

All of these eminently reasonable opinions can be understood when the essence of modern science is seen to consist in its adoption of particular physicalistic principles, as is suggested in the present book.

8. THE PHYSICAL VS. THE SPIRITUAL

As has been advanced in this book, and is supported by the considerations of the present appendix, modern science is an epistemological endeavour whose categories are limited to what is physical. Such subjects as parapsychology might be compared with biology in this regard. Biology is the modern-scientific discipline concerned with *life*, while parapsychology concerns the *spirit*. But since life, like the spirit, lies beyond the categories of modern science, why should biology be a part of science while parapsychology is not? The answer is that biology is only concerned with the *physical aspects* of

---

[62] Stokes (1985), p. 395.
[63] The same applies with respect to life: cf. n. 67 below and accompanying text.
[64] Stokes (1985), pp. 384–385.
[65] Alcock (1985), p. 562.
[66] As cited in ibid., p. 559.

life; the notion of life itself is not a category of modern biology.[67] More generally we could say that the only extent to which modern science is capable of dealing with non-physical substances is in terms of their physical aspects or manifestations. Thus, for example, self-willed action can only be the subject of modern-scientific research in the form of physical behaviour.

Conceiving of the situation in terms of *paradigms*[68] leads to the thought that while the core paradigm of modern science is *physicalistic*, it may be fruitful to think of magic, parapsychology and so on as tending to cohere about a core paradigm which is *spiritualistic*, and that the conceptual move from one to the other constitutes a *paradigm shift.*[69] Given this, we should then be inclined to say that just this distinction between the spiritual and the physical, manifest in philosophy for example as the mind-body problem, constitutes the most important epistemological distinction we humans have yet to make.[70] From this one might want to go on to say that it is thus only the physical that can be investigated *scientifically*, whether it be by modern science or some other form of science, and that this being the case constitutes *the fundamental limitation of science.* However, in part due to the honorific status attributed to the appellation "science," I am inclined to use the term in a broader sense so as to include investigation of the spiritual, as long as that investigation is *systematic.* On the philosophy of science being presented in this book it is not being assumed, as it is on the empiricist and Popperian views, that there is but one right way of going about acquiring

---

[67] Cf. *this volume*, pp. 149–150, 154, and 206: "But the nature of life itself, and how it differs from non-life or the physical, lies in principle beyond what the physicalistic categories of modern science could ever be able to handle."

[68] Cf. ibid., pp. 51–52&n., 56, 60, 70, 131, 133.

[69] As suggested by S. J. Tambiah (1990, p. 9) in the case of the transition from magical to modern-scientific thinking.

[70] Which is in keeping with my suggestion in Chapter 10 (p. 195) that the move to physicalistic explanation on the part of Thales initiated the greatest intellectual revolution in the history of humankind. Cf. also Price (1955), p. 217: "Rhine has correctly stated that 'Nothing in all the history of human thought – heliocentrism, evolution, relativity – has been more truly revolutionary or radically contradictory to contemporary thought than the results of the investigation of precognitive psi.'"

knowledge (or understanding) of reality, that being the *scientific*,[71] such that all forms of epistemological endeavour should conform to it. Here, rather, the possibility of there being different sciences and scientific methodologies is recognised, and no special value is attached to what is accepted as science today.[72]

The ontological approach adopted here, in which sets of principles function as paradigms for their respective subjects, also makes it clear that the difference between modern and non-modern science is not in all cases an either/or issue, but can involve the question of gravitation to a paradigm. This would explain why the scientific status of physics has never been questioned, despite its failure to meet the contiguity requirement with regard to any of the fundamental forces, and its apparent non-determinism and non-locality in quantum mechanics; the reason for its acceptance not being because of its methods, but because of its physicalistic categories. The present approach also explains why we can expect the (modern-)scientificity of the human sciences always to be questioned no matter what methodology they adopt. Further, it explains why, despite the fact that change on the Chinese worldview might be both deterministic and the result of the operation of contiguous causes, it is not accepted as a part of modern science, and why Chinese acupuncture is opposed by the American Medical Association, despite the technique's inductively demonstrated curative and anaesthetising qualities, the reason the Association gives being that acupuncture cannot as yet be explained mechanically.[73] And it explains why, despite the fact that the empirical methodology of parapsychology is strictly modern-scientific, the subject is even less accepted as a science than are the human sciences, as well as why elite mainstream scientists are those who evince the greatest scepticism regarding parapsychology, at the same time as they are the most inclined to cite a priori reasons for not accepting it.[74] The distinguishing feature in all of

---

[71] In this regard, cf. *this volume*, pp. 7–8, 61 and 72.

[72] Cf. the reference to Feyerabend's 'methodological anarchism' on p. 70 of ibid.

[73] In this regard cf. McClenon (1984), p. 78.

[74] Ibid., pp. 128 and 144. Cf. also Griffin (1996), p. 88: "Critical reflection about the paranormal is primarily important... for the same reason that it has been so difficult: because it challenges the modern paradigm ([i.e.] worldview)."

these cases is the various subjects' inclusion of categories of the spiritual; *this* is what keeps them from being sciences in the same sense as modern science.

9. CONCLUSION

Following the metaphysical approach of the present volume, not only can we demarcate modern science from the other epistemological endeavours treated above, but also distinguish it from such activities as mathematics, Western medicine and applied science, as well as from Platonic and Aristotelian science and common sense.[75] Not only this, but given the present approach we can also understand the hierarchy in which physics stands at the top, with, in order, chemistry, biology, the human sciences and investigations of the paranormal under it.

On the basis of the above reasoning then, we come to see that the distinction between modern science and other epistemological endeavours is well characterised when the subjects in question are considered in terms of their core principles. Most important in this respect is whether the substance of the practice in question is physical or not, and whether its causes operate only between physical entities.

---

[75] The last two of which receive a detailed treatment in Dilworth (2004).

APPENDIX III

REPLY TO CRITICISM

In this appendix I shall reply to many of the criticisms that have been made both of this volume as well as of the second and third editions of my earlier book, *Scientific Progress*.[1] I shall begin with the earlier book.

In the mid-1970s, when I began working on the topic of scientific progress, the incommensurability claims of Kuhn and Feyerabend, made in the 1960s, still constituted the central problem in the philosophy of science. The problem was that if scientific theories were incommensurable in the sense intended by Kuhn and Feyerabend, then the logical empiricist and Popperian views of scientific change (and of science more generally) would be seriously undermined: among other things, they would both lead to relativism due to their inability to handle the changes in meaning involved in the move from one of two incommensurable theories to the other.[2]

The response to Kuhn's and Feyerabend's claims on the part of most philosophers was to defend the 'received view' by trying to show the claims to be incoherent and thereby themselves to fail to avoid relativism – but in so doing these philosophers presupposed their own logico-linguistic approach, and the effect of their efforts was really only to show that the notion of incommensurability could not be captured on it.[3] In this regard, however, it is important to note that no one other than Feyerabend attempted to provide an alternative conception of scientific change capable of admitting incommensurability and at the same time avoiding relativism, and that his attempt was not successful.[4] This being the case may have contrib-

---

[1] I reply to criticisms of the first edition of *Scientific Progress* in Appendix II to its fourth edition.

[2] In this regard, cf. ibid., pp. 24 and 39.

[3] See ibid., pp. 76–78.

[4] Feyerabend's attempt took the form of his 'pragmatic theory of observation,' concerning which see ibid., p. 79&n.

uted to people's so readily taking both Kuhn and Feyerabend to be *advocating* relativism, and to Feyerabend's possibly later acceding to such an interpretation.[5]

At least partly as a result of this, a line has developed in philosophy of science in which the social aspects of science as treated by Kuhn are emphasised, while science is considered to be relativistic, while at the same time the mainstream view of the 1960s and earlier never solved its problem of incommensurability. Despite this and other major failings (as demonstrated in *Scientific Progress*), this view has continued to be the dominant line in contemporary philosophy of science. But *both* of these views are relativistic – the sociological view intentionally so, and the logico-linguistic view so due to its inability to solve the incommensurability problem, or more generally due to its inability to provide criteria for scientific progress that are applicable to actual science.

If one is to provide a non-relativistic conception of scientific change capable of handling incommensurability, as I attempt to do in *Scientific Progress*, the first question to be answered is what specifically is meant by saying that particular theories are incommensurable. I provide an answer to this question in Chapter 7 of the book, where I present both negative and positive senses of incommensurability, the former based on results I obtained employing the *Deductive Model* in the book's previous six chapters, and the latter as a first step towards introducing my own theory of science. Thus in the first regard I suggest that incommensurability implies that attempts to depict scientific progress based on the Deductive Model are inherently inadequate, and in the second that the gestalt-switch diagram illustrates an important aspect of incommensurability.

––––––––
[5] As I point out in *Scientific Progress* (p. 50), Kuhn was not advocating relativism; nor, at least in the 1960s, was Feyerabend. As I also point out there, "many of Kuhn's critics [and, as it turned out, many of his supporters!] presuppose, either wittingly or otherwise, that an account of science as a rational enterprise must take the form suggested by either the Deductive Model or some other formal construction . . . . [Thus] the relativism that arises in [contexts having to do with theory change] has been taken to imply a relativism in Kuhn's and Feyerabend's claims, [whereas] it is rather the Popperian conception that leads to relativism." (pp. 50–51, 53). Concerning rationality and science, see also *this volume*, p. 67, n. 32 and accompanying text.

As regards my own theory, the *Perspectivist conception*, which is introduced via the *Gestalt Model* and developed in the remainder of the book, it is to be noted that, as I remarked already in the first edition, "though the present view has gained much inspiration from the respective works of Kuhn and Feyerabend, it is not being presented as a direct reconstruction of the views of either of them."[6] The Perspectivist conception of science constitutes a philosophical theory that is to be judged on its own merits.

So *Scientific Progress* is devoted to providing a theory capable of both accommodating Kuhn's and Feyerabend's notions of incommensurability (which include the idea of theory conflict but not the Deductive Model) and at the same time avoiding relativism. In the words of Ricardo Gómez: "Dilworth attempts to show that his approach avoids not only the criticisms which Kuhn and Feyerabend level at the received view, but also the charges of relativism which in turn have been levelled at Kuhn and Feyerabend themselves." [7]

1. THE DEDUCTIVE MODEL

The Deductive Model of science is an abstraction I made on the basis of studying the writings of Hempel, Popper and others of that generation. This model lies at the core of my explanation of logical empiricism and Popperianism. What is important as regards the model is that, as I try to demonstrate in the book, both views presuppose it, such that all their central concepts can be expressed in terms of it, and all of their problems (including that of relativism) emanate from its use.

David Oderberg, in his review of *Scientific Progress*, suggests that most of my objections to logical empiricism and Popperianism "stem from the model to which these views are wedded," i.e. from the Deductive Model, while at the same time none of my criticisms are original.[8] In this regard I would first suggest not that *most* of my objections to empiricism and Popperianism stem from the Deductive

---

[6] *Scientific Progress*, p. 67.
[7] Gómez (1992), p. 264.
[8] Oderberg (1997), p. 188.

Model, but that *all* of them do. And I believe that a number of my criticisms are quite original. These include that the logical empiricist and Popperian views are formally identical (pp. 11ff.); that it is impossible for Popper to distinguish science from non-science on the basis of falsifiability once he claims, as he does in criticism of the empiricists, that no empirical statement is verifiable (p. 14); that his notion of corroboration is formally identical to the logical empiricist notion of confirmation, and thereby commits him to induction (p. 17) – for which he also criticises the empiricists; and that he provides no notion of content applicable to contradicting 'theories' (p. 34). To this I would like to add that, so far as I know, not only are all of these criticisms original, but they are also *correct*.

But much more important as regards the question of originality is my reconstructing the *whole* of logical empiricism (including my provision of the empiricists with a formal conception of scientific progress)[9] *and* Popperianism in terms of the Deductive Model, a reconstruction showing that both views *are* in fact "wedded" to the model. Oderberg apparently does not realise that this model (as distinct from the deductive-nomological model of explanation) was presented explicitly for the first time in *Scientific Progress* in an attempt to explain these two views, and that the first six chapters of the book are specifically devoted to this effort, which is to function as a lead-up to my own theory.

Gómez appreciates this however, as is clear from his commenting on the excellence of my attempt "to show that those shortcomings and problems [of the logical empiricists and Popperians] are mainly grounded on the shortcomings and problems of a commonly shared deductive model." Here I would claim that what is "excellent" regarding my attempt is that it is successful, as is evidenced, for example, by Oderberg's assuming that the empiricist and Popperian views *are* in fact "wedded" to the model.[10] Furthermore, I would suggest that my treatment of the views of the logical empiricists and Popperians in *Scientific Progress* is *definitive*.

---

[9] *Scientific Progress*, pp. 20–21; quote following, Gómez (1992), p. 265.

[10] Though reviewers of the second and third editions of *Scientific Progress* apparently appreciate that both logical empiricism and Popperianism do rely on the Deductive Model, some critics of the first edition questioned this.

Since the Deductive Model may be confused with the deductive-nomological (D-N) model of explanation, perhaps the difference between the two should be clarified. The D-N model can be seen as constituting a particular application of the Deductive Model, namely to scientific explanation and prediction. The Deductive Model itself is much more general, and also includes the logical empiricist notions of verifiability, induction, confirmation, progressive theory change, theoretical terms, and correspondence rules; and the Popperian notions of falsifiability, basic statement, background knowledge, corroboration, severity of tests, theory conflict, theory content, probability of a theory, testability, verisimilitude, and the whole of Lakatos' sophisticated methodological falsificationism.[11]

What the underlying identity of the logical empiricist and Popperian views means, among other things, is not only that they are not as different as Popper would have had us believe when he claimed to have 'killed' logical empiricism, but more importantly that both views are *dependent* on the Deductive Model in their respective depictions of the scientific enterprise. The model constitutes the *conceptual paradigm* for both of them, and any notions they advance in attempting to explain the nature of science must be expressible in terms of it.

2. THE PERSPECTIVIST CONCEPTION

To my mind, none of my current or earlier critics have appreciated how important the common presupposition of the Deductive Model on the part of the logical empiricists and Popperians is as regards their respective characterisations of science in general, nor as regards the incommensurability of scientific theories in particular. I might also say that few if any mainstream philosophers of science – including my present reviewers – today realise the extent to which they themselves are still ensconced in this logico-linguistic way of thinking.

---

[11] In this last regard, as also appreciated by Gómez: "Dilworth correctly emphasizes Lakatos' endorsement of many of Popper's views and, consequently, Lakatos' acceptance of the Deductive Model." (ibid., p. 265).

This may account not only for their general failure to fully appreciate the critical part of *Scientific Progress*, but for their failure to appreciate its positive part as well. Where Oderberg saw little novel in my work with the Deductive Model, Gert König believes the Perspectivist conception itself to be a *summary* of Hanson's, Kuhn's and Feyerabend's "investigations of the fundamental perspectival shifts of science."[12] However, not only is the Perspectivist conception not a summary of anyone else's work, but the notion of perspectival shifts *originated* with it.

Similarly, Hanne Andersen considers the Perspectivist conception to be "an interesting elaboration of ideas which were only briefly introduced by Hanson, Kuhn and Feyerabend."[13] Here Andersen apparently does not realise that it was not my intention to provide an elaboration of other people's ideas, but rather to provide an independent theory of scientific progress which took account of their various claims; nor does he seem to realise that the fundamental ideas of the Perspectivist conception are not to be found in the work of Hanson, Kuhn or Feyerabend.

Andersen also criticises me for not referring to Kuhn's later work and for not discussing the work of such logico-linguists as Scheffler, Putnam and Kitcher, who are supposed also to have had the idea that "differing conceptual perspectives can have the same reference."[14] But in the first case, my interest is not in Kuhn's work *per se*; and in the second, once again, "conceptual perspective" is a technical term that presupposes the Perspectivist conception. If the authors Andersen mentions are in fact speaking about conceptual perspectives (which I doubt), then they are indebted to *Scientific Progress* for the notion.

The Gestalt Model

Regarding the Gestalt Model, Gómez says: "Dilworth believes that to perceive something in a particular way, e.g. as a rabbit (per-

---

[12] König (1989), p. 371.
[13] Andersen (1997), p. 265.
[14] Ibid., p. 266; *Scientific Progress*, p. 80. Cf. also the treatment of Putnam on pp. 47–48 of *this volume.*

ceptual perspective) may be considered to involve the application of a certain concept (rabbit) to it."[15]

This remark by Gómez is an instance of a form of comment on my work that often recurs and which is ill-founded, namely that of taking me to be making a factual claim when I am presenting a theory,[16] and suggests that those making such comments are unaware of the difference. Presenting a theory, as I see it, involves stipulating certain things regarding the categories or concepts used in the theory, in order to indicate more precisely what they are, which at the same time should give the theory structure. In the present case, what I am doing is *suggesting* that to perceive something e.g. as a rabbit may be considered to involve the application of a certain concept to the thing (independently of whether I believe it does or not). This suggestion is part of the presentation the Gestalt Model, which, in *Scientific Progress*, is itself part of the presentation of the Perspectivist conception of science.

Identification of the Intended Domain

A number of reviewers have had difficulty with my notion of an intended domain. Andersen, for example, suggests that my *claim* that "differing conceptual perspectives can have the same reference" requires a theory of reference;[17] Oderberg wonders how the scientist is to *know* what it is his theory is supposed (by him) to apply to (!?);[18] and Gómez demands a better account of how the intended domain is identified.[19]

---

[15] Gómez (1992), p. 267.

[16] Gómez makes the same sort of mistake earlier in his review when he says that I *believe* that each intended domain is neutral with respect to the particular views under consideration. Similarly, Andersen takes me to be *claiming* that differing conceptual perspectives can have the same reference (1997, p. 266), while Oderberg takes me to be stating as a matter of fact that that to which a model is applied is determined by the intention of the individual applying it (1997, p. 193), when in all of these cases what I am doing is presenting part of my theory. In this regard see also *this volume*, p. 165, n. 30.

[17] Andersen (1997), p. 266 – see also previous note; *Scientific Progress*, p. 80.

[18] Oderberg (1997), p. 193.

[19] Gómez (1992), p. 267.

I believe, however, that I have made it quite clear in *Scientific Progress* what the intended domain of a perspective is, and how it is to be determined – both from a subjective or first-person and from an objective or third-person point of view; and I believe further that the problem these writers find with my clarification is not because it is flawed or incomplete, but because their inability to relinquish the logico-analytic conception of reference has prevented them from understanding it.

The core of the clarification of the nature of the intended domain and how it is to be determined on the Perspectivist conception is as follows:

[W]e should say that when a scientist moves from one of two incommensurable theories to the other, he can still very well *intend* that both theories apply to the same states of affairs, even if they characterise those states of affairs in essentially different ways. Furthermore, on the basis of certain criteria (such as the performance of the same sorts of operations), we, as onlookers, can often judge that the scientist is treating of one and the same aspect of reality in his respective applications of the two incommensurable theories.[20]

Here we have a characterisation from a *subjective* point of view: the intended domain is determined by the *intention* of the theorist (if *he* doesn't determine what he's applying his theory to, who or what does?). And we have a characterisation from an *objective* point of view: given that we are not in the position simply to *ask*, our criteria for determining e.g. whether two theories have the same intended domain depend on such intersubjective criteria as the nature of the operations involved in their respective applications.

At various points I specify what I mean in greater detail. For example I link the notion to the Gestalt Model:

[D]iffering conceptual perspectives can have the same *reference*. The use of this term is to suggest the idea that a conceptual perspective, in being an applied concept or system of concepts, in a sense 'points' in a certain direction. And the direction in which it 'points' is in turn dependent upon the *intention* of the person applying the concept or conceptual framework, and not on the concept itself. Thus, as has been suggested in the previous chap-

---

[20] *Scientific Progress*, p. 79; next quote, p. 80.

ter, the 'duck' and 'rabbit' concepts of the Gestalt Model might each be intended to apply to a different figure, and consequently be said to be given different references.

And I provide an example from science in explicating the kinetic theory of gases:

[T]he intended domain of a scientific theory is to be thought of as encompassing all of the empirical states of affairs to which it is intended that its model be applied. Thus we see that the reference given van der Waals' model is broader than that given the ideal gas model, in that it is intended to be applied not only to substances in their purely gaseous form, but also to such substances when they undergo a change of state.[21]

Of course one can attempt to apply the notion to more cases, and perhaps develop it further, but as regards the philosophical question of what I mean by the intended domain of a theory and how it is to be delineated, it would seem to me that what I have provided should suffice.

Categories and Perspectival Incompatibility

On the Perspectivist conception, if scientific theories involve the application of different predicates from the same category to the same intended domain (at the same logical time), then they should be *perspectivally incompatible*.[22] Perspectival incompatibility is to be independent of whether the competing perspectives suggest the same or different results. This idea is clearly exemplified in the Gestalt Model in the form of the cube gestalt-switch (the Necker cube): you can see the cube as though you were looking at it from either above or below, but the differences of perspective are not manifest empirically. Though such clear exemplifications may be more difficult to find in the context of science, the Tychonic and Copernican conceptions of the solar system constitute an instance of where two models conflict even though they suggest the same results.[23]

---

[21] Ibid., p. 95.

[22] Cf. ibid., p. 84.

[23] Other examples include the Machian vs. Newtonian conceptions of space (ibid., p. 200), the eccentric vs. epicycle hypotheses regarding the motions of the planets (see Heath, 1913, p. 266), and a thought-experiment involving perfectly circular

More than this, however, it is also suggested in *Scientific Progress* that in at least one area in physics (gas theory) the categories – space, time, mass, and so on – involved in different perspectives are *quantified*. Such quantified categories I call *parameters*. In the case of parameters, the predicates or concepts falling under them are rational number *values*; and perspectival incompatibility should arise between gas theories only if they respectively involve the application of the same parameter(s) but with different values.[24]

In this regard Andersen raises an interesting point:

> But if theories conflict only if they involve different concepts from the same category, and different concepts are understood by the 'quantified categories' view as different values, then theories can only conflict if they involve the same parameter but predict different values.[25]

I believe the key to meeting Andersen's point may lie in the distinction I draw in *Scientific Progress* between parameters quite generally and *measurable* parameters,[26] a distinction I develop further in the present volume. In this development I call measurable parameters *magnitudes*, suggesting that "theoretical laws may be expressed by equations; and while such equations must depict relations between quantities, given the hypothetical nature of theoretical models they need not depict relations between magnitudes."[27] Thus there can exist quantitative differences on the theoretical level without those differences being manifest in what is measurable on the empirical level.

We see a similar difference between mensural and non-mensural levels in the Tycho/Copernicus case. Though all the categories in the two models are (potentially) quantitative, the differences between

---

orbits about the sun, in the one case their being conceived of as due to a gravitational attraction, and in the other as due to circular motion being natural (*à la* Aristotle).

[24] Cf. *Scientific Progress*, pp. 89 and 97.

[25] Andersen (1997), p. 267.

[26] *Scientific Progress*, p. 98.

[27] *This volume*, p. 114; I also say that, "due to the possible inaccessibility of theoretical mechanisms, some of the properties represented in the theoretical equations, while being quantities, may not be magnitudes." (p. 116). As regards the hypothetical nature of theories, cf. ibid., esp. Chapter 4, Section 4.

the two views – their different conceptions of the motion and rest of the sun and the earth – are nevertheless not *measurable*, and thus cannot be manifest in their results. In this case the difference concerns the category of space, the one perspective taking the earth to be motionless in space and the other taking the sun to be so. But this difference, while physically conceivable, has no influence on the empirical level since space itself cannot here be quantified in such a way as to link it to a relevant magnitude. While space is quantifiable in certain mensural respects – as regards distance etc. – in the present example it is not quantifiable with regard to absolute motion, and so the perspectival incompatibility is not manifest on the empirical level.

What in this regard is important about my notion of perspectival incompatibility, however, is that it suggests that the essence of theory conflict lies in the conceptual cores of the theories, and not in whatever empirical results as may be derived from them. Competing theories constitute mutually exclusive ways of conceiving of reality which conflict with each other as *wholes*. Nevertheless, it is not to be denied that Andersen's comment points to certain aspects of the Perspectivist conception that might be further developed. Such a development could involve, for example, considering whether *all* categories expressed in *all* physical theories *must* be quantifiable (and if so, why), whether some category or categories (e.g. that of cause) are different than others in this or related respects, and under exactly which conditions differences of category on the theoretical level should or should not lead to different empirical results.

3. PRINCIPLES

Moving on to the present book, its point, as expressed in its subtitle, is to provide an *account* of *modern science* in terms of three basic notions, namely *principles*, *laws* and *theories*. Its point is *not* to account for the fundamental principles of modern science and their history, as Andersen takes it to be.[28] Nor is it correct to say, as

---

[28] Andersen (1997), p. 268.

F. Weinert does,[29] that I see the major value of the book to lie in my *analysis* of such principles, the implication being that I simply assume everyone to take them to exist and to be the ones I indicate. Though I do analyse these principles, they are the principles that *I* suggest to lie at the core of science, and my analysis is of interest only to the extent that it helps demonstrate how modern science as a whole can be understood in terms of them.

Principles are however the most important of the three basic notions, since in the book the whole of science is taken to presuppose them. And the actual principles from which science is to emanate are specified, just as Kant and Whewell each specified what they took the fundamental principles of science to be.[30]

After Chapter 1, reference is made to work in the philosophy of science only when this work can help clarify the nature of science, or when it anticipates some aspect of the view I am presenting. In any case, there is a shift from dealing with a problem in the philosophy of science in *Scientific Progress*, to dealing with the nature of science itself in this volume.

Something the two books have in common, however, is that the content of each may in fact be broader than just to cover modern science. The Perspectivist conception of *Scientific Progress*, particularly as applied through the Gestalt Model, has a potentially much wider application than just to theory change in modern science,[31] as does the Gestalt Model, which, for example, I have applied by itself to the issue of identity and reference in the philosophy of language.[32] And, as far as the present book is concerned, the principles dealt

---

[29] Weinert (1997), p. 330.

[30] In Kant (1786) and Whewell (1847), respectively.

[31] As regards modern science itself, one can also imagine acquiring insight into e.g. shifts between systems of co-ordinates in relativity theory, and the particle/wave dualism of quantum mechanics, by analysing them in terms of the Gestalt Model and/or the Perspectivist conception. Moving beyond modern science, "there is no reason why we could not in a different setting consider, for example, such geometrical theories as those of Euclid, Lobachewsky, and Riemann each to constitute a conceptual perspective being applied to geometrical space." (*Scientific Progress*, p. 69). The Perspectivist conception was also applied, for example, to the issue of *sustainable development* in Appendix VI of the third edition of *Scientific Progress*.

[32] In my (1986).

with in it might, in their more abstract form, be much more universal, such that perhaps all epistemological endeavours could profitably be analysed in terms of them.[33]

To succeed in providing a general philosophical account of modern science requires, among other things, showing how the central notions in the account are manifest in central aspects of science, and thereby indicating why those aspects are related to one another in the particular way that they are. The central notions in the present account being those of principles, laws and theories, the explanation includes showing in detail how, for example, in modern science empirical *laws* are explained by *theories*, which perform this function by indicating how the laws are nothing other than particular manifestations of the *principles*.

With regard to these distinctions, Louk Fleischhacker contributes some interesting thoughts:

The starting point of this account is the observation that three levels of thought can be distinguished within science: that concerned with (empirical) *laws*, that concerned with (originally speculative) *theories*, and that concerned with (metaphysical) *principles*. If one were to take these modes of thinking independently of one another, each might lead to its own conception of science. The first would tend towards empiricism, the second towards realism, and the third towards rationalism. Dilworth, however, sees them as an integrated whole, in which thought concerning principles determines the nature of thought regarding laws and theories.[34]

The Principles of Science are Relatively *A Priori*

Perhaps the two most important features of principles with respect to how they are conceived in the present volume and how they have been conceived in the past is that here they constitute the *core* of the scientific enterprise, i.e. together they function as a *paradigm*; and, as is related to this, they are merely *assumed* for the purposes of doing science, and are not taken as being *known* to be true.[35] As

---

[33] In this regard, see the preceding appendix and Dilworth (2004).

[34] Fleischhacker (2002), p. 324.

[35] These aspects of the present approach are very important in distinguishing it from other approaches such as those of Kant and Whewell, where the a priori underpinnings are to constitute an absolute foundation.

noted by Chris Eliasmith,[36] this means, among other things, that they are not conceived of as necessary truths, nor as axioms from which other true statements can be formally deduced, but are rather conceptions that are subject to revision.

Oderberg raises an interesting question with regard to principles when he asks: "Are the principles 'adopted,' or are they *there*, whether or not scientists recognize their reliance on them?" This question has rather general implications concerning the doing of philosophy. The principles are adopted, but probably not consciously. Nor can they be said to *exist* in any real sense even when they are unconsciously adopted, other than perhaps in the backs of the minds of those adopting them. From one point of view they are nothing other than the referents of a particular abstract conceptual scheme applied to the situation after the fact in order to make as much sense of it as possible.[37]

Alexander Bird, for his part, wants to know what the fundamental principles of modern science explain.[38] But principles do not explain anything; they constitute the core from which explanations emanate; they are what is *presupposed* in any explanation. Bird further suggests that the fact that modern science presupposes the uniformity principle does little to characterise the enterprise, except to the extent that if scientists did *not* presuppose it, they would not bother going to work. Apart from the fact that, as I show in detail in Chapter 3, the whole of the empirical aspect of science rests on the principle, even if the characterisation of its effects were limited to suggesting that without it scientists would not bother going to work, this in itself would constitute important information concerning the nature of science.

---

[36] Eliasmith (1998), p. 657; quote following, Oderberg (1997), p. 193.

[37] Cf. *this volume*, p. 71–72; it may be of some interest to note that this line of thought also applies to the Deductive Model. In the present volume what are being referred to as ontological principles are *conceptual*; the distinction between conceptual principles and their potential correspondents in reality (*real* principles) is taken up in my book *Simplicity*, presently in preparation.

[38] Bird (1997), p. 285.

The Primacy of Ontological as versus Methodological Principles

In this book the thesis is presented that the nature of science can be better understood by taking it to be based on particular *ontological* principles than in any other way, including taking it to be based on particular *methodological* principles (such as those of verifiability or falsifiability). In this regard Joel Katzav suggests that:

many philosophers who hold that it is method which defines science would not . . . wish to deny that in general the theories of modern science have been formulated with something akin to the metaphysical principles Dilworth spells out in mind.[39]

In response to this I should first say that, as regards the question of defining science in terms of method, as I have demonstrated elsewhere, efforts to do so have been quite unsuccessful.[40] The next question concerns how best to *explain* science, so if 'methodologists' were to accept my explanation in terms of ontological principles this would only be to the good, particularly considering that they themselves have been unable to explain science in terms of method. However, if we grant, as Katzav suggests, that 'methodologists' would accept something similar to the ontological principles I indicate, it may then be asked with what right they should do so. What is it in their methodological principles, whether they be of verifiability or falsifiability or whatever, that should allow them to say, e.g., that some form of the principle of the uniformity of nature holds for physical reality as conceived in science? As implied by Hume's arguments concerning induction, the view that reality is uniform demands an act of faith. In other words it demands the acceptance of a particular ontological configuration *before* the methodology is put into effect.

Katzav says further that,

if the metaphysical principles used by a community were to differ radically enough from those of modern science, Dilworth would not apply the term modern science to describe the practices of that community, whereas those

---

[39] Katzav (1997), p. 316; next quote, ibid.
[40] In this regard see *this volume*, p. 241&n.

who take method to characterise modern science might still wish to do so. However, this difference is merely a matter of definition, and not a dispute over matters of fact.

This strikes me as a very strange thing to say. Philosophical disputes are never over matters of fact; and such "matters of definition" as Katzav refers to determine how we conceive of reality. The question of whether it is better to conceive of modern science as based on ontological or methodological principles should be of central importance to the philosophy of science. On Katzav's way of thinking, *any* result obtained in the philosophy of science ought to be classed as 'merely a matter of definition.' Regarding the distinction between science and non-science in particular, great emphasis has been laid in the academic world at large on the attaining of scientific status. The whole thrust of the empiricist and Popperian methodologies, with their respective distinctions between science and pseudo-science, was to indicate what should and should not be considered to be a scientific form of investigation. If one philosophical approach says that a certain form of inquiry is (modern-)scientific while another does not, the matter is certainly of greater importance than simply to be termed one of definition.

Refined Principles

Principles are of two sorts, *fundamental*, which constitute the core of the whole of science, and *refined*, which are derived from the fundamental principles, and, among other things, define various scientific disciplines.

Hanne Andersen suggests that the relation between fundamental principles and refined principles could be made clearer.[41] Here I shall attempt to do this in the case of the principle of causality and Newton's first law.

Newton's first law says that a body will continue in its state of rest or non-accelerated rectilinear motion unless acted upon by a force. The principle of causality says that change is caused. What is the relation between these two locutions?

---

[41] Andersen (1997), p. 269.

Following Whewell,[42] in this book the notion of *force* in Newton's laws is taken to be the expression of the more universal notion of *cause*: force is a *type* of cause; and *change of state* is the expression of the more universal notion of *change*: change of state is a *type* of change.[43]

From one point of view, the first law is a *definition* of "cause" and "change" *for Newtonian mechanics*, where *cause* is defined as *force*, and *change* is defined as *change of state. Given* these 'definitions,' in the context of Newtonian mechanics the first law and the principle of causality are *identical.*

Note that in this way the application of Newton's first law is restricted to the bodies to which Newton's mechanics applies, whereas the principle of causality is not. The notion of physical force is a particular instance of the notion of cause, and not the other way round. In this way Newton's first law is *derived* or *drawn* from the principle of causality.[44] It indicates a particular domain in which the principle is to hold, and is thus also a *specification* of the principle.

The question of the extent to which Newton's laws ought to be considered either empirical or a priori is taken up by Weinert. He suggests that on my characterisation of empirical laws at least New-

---

[42] Cf. *this volume*, pp. 11, 29n., 62n. and 64n. With reference to Newton's first law and the principle of causality in particular, Whewell says: "If we call to mind the axioms which we formerly stated, as containing the most important conditions involved in the idea of Cause, it will be seen that our conviction in this case depends upon the first axiom of Causation, that nothing can happen without a cause. Every change in the velocity of the moving body must have a cause; and if the change can, in any manner, be referred to the presence of other bodies, these are said to exert *force* upon the moving body: and the conception of force is thus evolved from the general idea of cause. *Force is any cause which has motion, or change of motion, for its effect.*" (1847), Part 1, p. 217.

Newton himself implies that force is a species of cause, where he says that "[t]he causes by which true and relative motions are distinguished . . . are the forces impressed upon bodies to generate motion" (1687, p. 10); and among modern writers it is taken for granted e.g. by Ellis in his (1965): cf. esp. pp. 39–40, 47 and 53–54.

[43] Cf. *this volume*, p. 64: "[C]hange consists in change of state (rather than change of position) due to the presupposition of the substantiality of motion."

[44] Cf. ibid., p. 62: "[S]uch refinement does not consist in the refined principles' being formally *deduced* from the more fundamental principles, but rather in their being *drawn* from them in such a way as to allow the fundamental principles to be *applied* in particular cases."

ton's second and third laws ought to be considered empirical, while I claim that Newton's laws ought to be seen as being principles.[45] The point is, however, as I say in the book, "these roles do not exclude one another, and so such questions as to whether principles function in one of the ways treated here rather than another do not arise." And in this general regard I point out a weakness in the logico-linguistic approach: "many questions in analytic philosophy concern the status of particular expressions without at all considering that they can have a different status in different contexts."[46] Refined principles can and do function in different ways in different contexts.[47] In fact part of the very refining of fundamental metaphysical principles consists in making them applicable to reality,[48] in the case of modern empirical science, ultimately to mensural reality.

4. REALISM VS. EMPIRICISM

In the introductory chapter of this book the historical debate over empiricism and realism in the philosophy of science is presented. This is done in order to afford a detailed depiction of a topical, complex issue in the philosophy of science so as subsequently to be able to show how it can be handled by my metaphysics of science. The content of the book itself stands above such issues. As I say in the chapter on principles, "[no] prescriptive stance [is] being assumed with regard to the empiricism/realism debate. ... Rather, the main aim of the present work is to capture the essence of modern science,

---

[45] Weinert (1997), p. 331; *this volume*, p. 65. Quote following, ibid., pp. 65–66.

[46] Ibid., p. 66, n. 29. In this regard, cf. also my remarks concerning 'context-blindness' in Dilworth (1992), pp. 207–210.

[47] In this regard cf. *Scientific Progress*, p. 134: "The question often arises in considerations of Newton's theory as to whether his axioms or laws of motion ought best to be taken as definitions (or a priori truths), or as empirical laws, and the conclusion usually drawn is that they can function in both sorts of ways. This conclusion ... is easily accommodated on the present view."

As expressed by Ellis: "Newton's second law of motion thus has a variety of different roles. ... To suppose that [it] must have a unique role that we can describe generally and call the [epistemo]logical status is an unfounded and unjustifiable supposition." (1965, p. 61). In this regard cf. also Hanson (1958), pp. 97ff.

[48] Cf. n. 44 above.

and in so doing indicate the positions of empiricism and realism with regard to it."[49] As it turns out, my metaphysics of science vindicates realism at the expense of empiricism. But things could have been the other way round, such that it favoured empiricism over realism. What is important is that the metaphysics presented in the book can *resolve* this issue; that it shows one or the other of these views to be more in keeping with the nature of modern science is incidental. Some of my reviewers have appreciated this fact, and others have not.

Ian Hinckfuss grasps the idea where he says:

Dilworth aims to give a description of modern science and use it to resolve the main philosophical issues in the philosophy of science, including the problem of induction, the nature of scientific reduction, the problem of natural kinds, and the debate between what he calls 'empiricism' and 'realism.'[50]

And Weinert also appreciates it:

Dilworth attempts to show how the principles are related to the other two pillars of science, laws and theories. In the process of clarifying these relationships, Dilworth also hopes to shed light on the empiricism/realism debate in current philosophy of science.

Some other authors, however, have misunderstood my intentions, and have taken me be a realist,[51] or to be presupposing,[52] defending,[53] or arguing for[54] realism. These reviewers are working on a level where the empiricism-realism question is as deep as it gets. This is how the majority of modern commentators deal with the issue, not realising that in most cases they are at the same time presupposing an empiricist conception of science.

With regard to my treatment of this issue, Eliasmith suggests that my choice of principles stacks the deck in favour of realism: "Hav-

---

[49] *This volume*, p. 52. It may be noted that I do not presuppose or defend either empiricism or realism in *Scientific Progress* either. There I say, for example, that "the Perspectivist view may be seen to be in keeping with both realist and instrumentalist conceptions of scientific theory." (p. 166, first three editions).

[50] Hinckfuss (1998), p. 130; next quote, Weinert (1997), p. 331.

[51] Bird (1997), 284.

[52] Eliasmith (1998), p. 658.

[53] Oderberg (1997), p. 193.

[54] Katzav (1997), p. 316; quote following, Eliasmith (1998), p. 658.

ing built a realist position into these core principles of science, it is hardly surprising that Dilworth determines that science is a realist project." But what is key here is that my metaphysics of science can handle all of the central aspects of science, including the role played by theories (thereby e.g. solving the problem of theoretical terms), as well as function in other ways as well, such as by affording a framework for the comparison of modern science with other epistemological activities, as in the previous appendix. If Eliasmith believes some other choice of principles could do this and at the same time vindicate *empiricism*, then he should indicate just what those principles are, and how they accomplish that end.

Katzav, while he on the one hand sees me as arguing for realism, nevertheless captures the spirit of what I am suggesting where he says:

Knowledge of empirical laws and the attempt to gain understanding by explaining them are...the two basic aims of science. Moreover, it is these aims that give science its realist flavour. Conceived of as above, science is [realist] in the sense that (1) the empirical laws relate quantities and these are not directly observable, and (2) making sense of facts involves conceiving of a particular ontology which *might* ground them. Indeed...not only is science [realist] but also it *ought* to be so, since the attempt to gain understanding via the principles of science is essential to the scientific enterprise.[55]

5. UNDERSTANDING VS. KNOWLEDGE

I considered the separation of scientific *knowledge* from scientific *understanding* to be very important already in *Scientific Progress*;[56] and the results of my investigations in preparing the present volume only strengthened this conviction. As noted by Eliasmith, on the view advanced in this volume, "Laws provide scientific *knowledge* (and are [conceptually] prior to theories), whereas theories, by linking laws to principles, provide scientific *understanding*."[57]

---

[55] Katzav (1997), p. 315.
[56] Cf. *Scientific Progress*, pp. 174, 182–183 and 207.
[57] Eliasmith (1998), p. 657.

I go on in this volume to indicate how the understanding provided by theories need not be a *correct* understanding.[58] When theories are first put forward they are only hypothetical, and thus so is the understanding they afford. This of course is not to say that they may not sooner or later be determined to be essentially correct (or mistaken), in which case we can then speak of them as providing a correct (or faulty) understanding. Part of this idea is captured by Katzav, where he says:

Dilworth, it should be noted, leaves open the question of whether we can actually have knowledge of the entities referred to by scientific explanations. [T]he attainment of understanding, one of the two central aims of science, does not require the attainment of knowledge of a transcendent realm.[59]

Thus, I find it difficult to understand Hinckfuss' saying: "Dilworth correctly draws a distinction between knowledge and understanding, but fails to allow that understanding may be a sub-species of knowledge." Considering that I go to great lengths to indicate how understanding *differs* from knowledge, including that it can be mistaken while knowledge cannot, this comment seems quite out of place.

6. THE PTL MODEL OF SCIENTIFIC EXPLANATION

The determination of the nature of scientific explanation must be a central aspect of any philosophy of science in which the notion is admitted. In this book I devote the whole of Chapter 5 to this question, where I present the Principle-Theory-Law (PTL) model of explanation. This conception is rather well described by Bird:

The job of the [theoretical] scientist is to explain [empirical laws], that is, to show how they are manifestations of the discipline's basic principles. This is what is achieved by the theoretical model. One instance might be the following. Boyle's law is an empirical law – an observed regularity. The principles of our discipline are Newton's laws of motion. Our theoretical model has a 'substantial' element – the ontological bit which tells us what sorts of stuff

---

[58] *This volume*, p. 105.
[59] Katzav (1997), p. 315; next quote, Hinckfuss, p. 132.

we are dealing with, molecules in this case. It employs the [refined] principles, Newton's laws, working on the substance, the molecules, and by relating magnitudes at the theoretical level with those at the observable level,
e.g. mean kinetic energy and temperature, it is able to derive the empirical
law.[60]

One criticism of the PTL model is that of Weinert, who says:

[I]t is hard to see why the subsumption of a particular fact or a regularity
under a fundamental principle should be an explanation of the fact or the
regularity. ... To be told that Snell's law or Ohm's law are but manifestations
of the principle of uniformity is not to be told why Snell's or Ohm's law take
their particular form.

Weinert appears to be missing a few of points here. First, as I noted
in response to Bird above (p. 283), the role of principles is not to
explain, but to constitute the basis or core from which explanations
are made; that empirical laws are the manifestation of the uniformity
principle is *presupposed*. Second, their *explanation* consists in showing *how* each results from the operation of the principle of *causality*,
normally mediated by the principle of *substance*. Third, as should be
impossible to miss due to the very name of the Principle-Theory-Law
model, such an explanation requires the construction of a *theory* in
which the principles are clearly manifest and from which the laws in
question can be derived. With regard to Weinert's example, it is not
the uniformity principle that indicates why Snell's law and Ohm's law
take the particular form that they do, but Maxwell's electrodynamic
theory.

Weinert goes on to suggest that "the PTL model is still a subsumption
account of explanation, but one in which the appeal to causal laws and
mechanisms is of primary importance."[61] As regards the latter point,
what is of primary importance are the principles being presupposed in
the explanation; it is the principles which may or may not lead to the
conceptions of causal laws and mechanisms. The existence of such
entities is not being *appealed to*, but *explained* – a point to which I
shall return below.

---

[60] Bird (1997), p. 285; next quote Weinert (1997), p. 333.
[61] Ibid.; next quote, Hinckfuss (1998), p. 131.

But as regards the first point, that the PTL model provides a subsumption account of explanation, while it may be true, it leads one to think that the account I provide involves the idea of *formal* subsumption, which it does not. Ian Hinckfuss goes further in this direction when he says:

The PTL model is supposed to explain how the theories explain how the empirical laws are a manifestation of the refined principles. Presumably they would do this only if the empirical laws were deducible from the principles and theories. [This] is in line with the D-N model.

What these authors apparently fail to realise is that no matter what model of scientific explanation one employs, what is explaining must, *in some sense*, subsume what is being explained.[62] This subsumption takes the form of logical deduction on the D-N model, but not on the PTL model. I have used the term "derivation" in this respect, and have shown in detail in Section 3 of Chapter 5 how it works – which is in a way that is clearly different from that of formal subsumption.

What is key here, however, is how the D-N and PTL models are expressions of fundamentally different ways of conceiving of science. The PTL model is the expression of how explanation is to be understood according to a particular conception of modern science which sees the enterprise as emanating from three particular ontological principles, with the principles being indicated. The D-N model, on the other hand, is essentially an expression of the view presented in Aristotle's logic (as developed by the Stoics), which the contents of this book (and of *Scientific Progress*) should show clearly to be conceptually far distant from the phenomenon of modern science (as it is from Aristotelian science, for that matter). It would appear that both Hinckfuss and Weinert are merely trying to understand my view in terms of their own logical-empiricist paradigm.

---

[62] As I say in another context, i.e. with regard to the Perspectivist conception, where the subsumption in question is also not formal: "in some sense, succeeding theories do subsume their rivals," and "[t]hough there is a sense in which the rabbit aspect both is superior to and subsumes the duck aspect, this superiority does not rest in the latter's being formally deducible from, or reducible to, the former." (*Scientific Progress*, pp. 49, 65).

Causal Mechanisms

Similarly to the claim that I am a realist, it has also been suggested that I *rely* on the notion of causal mechanisms, and that I am *endorsing* a causal conception of scientific explanation.[63] In reply I might begin by asking what it is that I am "relying on the notion of causal mechanisms" in order to do. I am certainly not relying on it in order to present my metaphysics of science. In fact I am not relying on it for anything. I am *explaining* why the notion is so important in modern science (as is apparent, for example, from the views of various non-committed commentators cited in Appendix II). What I am *relying on* in the present work is that the three conceptual principles I take to be at the heart of science actually are so. Given that they are, one's thinking is led to a conception of science in which the role of what are generally admitted to be causal mechanisms in science becomes clear. That there exist theories representing causal mechanisms is a *result* of the effect of the principles of modern science on the discipline as a whole. And what I am *endorsing* is not a causal conception of scientific explanation, but rather a particular way of seeing science, which is such that the commonly accepted idea that science attempts to provide causal explanations is both explained and shown to be correct.

7. QUANTUM MECHANICS

The idea that the principles of science are *relatively* a priori, as advanced in this book, is a new way of looking at the relation between (conceptual) principles and the epistemological enterprise (in this case, modern science) for which they are the principles. As mentioned above (p. 282), the relativity of the principles of science takes the form both of their constituting the *core* and not the *basis* of science, and in their being emendable given the results of scientific inquiry itself.

With regard to quantum mechanics in this context, Weinert says:

[Dilworth] affirms that revolutionary changes in 20th century science like

---

[63] Weinert (1997), p. 332.

relativity theory, quantum mechanics and chaos theory involve an alteration of epistemological principles rather than of ontological principles. These changes, he continues, do not involve ontological insights. The literature on the philosophy of quantum mechanics, at least, says otherwise. The Bell inequalities show that such fundamental classical notions as locality have no place in ontological interpretations of the quantum world. The quantum world seems to require a non-classical ontology.[64]

First, it should be pointed out that on my metaphysics of science chaos theory is not a part of modern science;[65] and second, that Weinert is mistaken in saying that I hold that revolutionary changes in twentieth century science involve an alteration of epistemological principles rather than of ontological principles. Regarding this latter point, I explicitly say on the contrary that: "Since the turn of the twentieth century however, the [ontological] principles of modern science in its core discipline of physics have come to be qualified in ever more drastic ways."[66] But I also say that when it comes to change of principles, the change is more radical when seen from an empiricist than from a realist point of view.[67] In other words, empiricists are more inclined than realists to see epistemological states of affairs, such as the inability to *detect* a causal link, or the inability to *determine* the simultaneous position and velocity of a particle, as implying that there *is* no causal link and that particles *as a matter of fact* do not have both a position and a velocity at the same time, while realists, on the other hand, are more inclined to leave such questions open, or to investigate the situation further.

Rather generally when it comes to the core principles of a discipline and their emendation, I suggest that even when an emendation

[64] Ibid., p. 334.

[65] *This volume*, pp. 189–190.

[66] Ibid., p. 130. Cf. also pp. 107, 205, and 69: "if... every possible way of theoretically reconciling the phenomena with the ontological principles in question appears blocked, the principles may themselves come to be questioned, and eventually be emended or perhaps even replaced."

[67] Ibid., p. 188. Cf. also p. 176, n. 6: "One here sees how the propensity interpretation is positivistic or empiricist in its orientation, since it takes limitations on the acquiring of empirical knowledge as being a direct indication of a characteristic of the world being investigated. Here, as elsewhere, the ontological collapses into the epistemological on the empiricist view."

has been generally accepted, it is not that the core principle in question has been discarded; rather, it has been *shelved*.[68] And, equally important, the categories in terms of which it has been framed are retained.[69]

Evidence suggesting the need to alter a particular principle in a scientific discipline is originally considered with scepticism. In the beginning it constitutes a Kuhnian puzzle; and if it persists it becomes a *problem* which, if it cannot be solved given a good deal of effort, may eventually lead to an emendation of the principle. And though less attention may thereafter be devoted to the problem, the door is always left open to its eventually being solved.

We see this quite clearly in the history of science with regard to the principle of contiguity. Though no mechanism has been found for gravity or the other fundamental forces, the discovery of such a mechanism, unlike Heisenberg's uncertainty principle or the Bell inequalities, would only be welcomed. This is how the core principles function, that is, as *ideals* which, while not always met in research, nevertheless tend to set the direction of that research. That contiguity still constitutes such an ideal is clear from the fact that scientists since Newton and up to the present have constantly been seeking a medium for the transmission of gravity.[70]

That this is also the case in the realm of quantum mechanics is manifest in the great attention that has been paid to Bell's inequalities and non-locality. This attention derives from the importance of contiguity and proximity to modern-scientific thought, and the fact that non-locality not only implies an absence of contiguity, but of proximity-dependent influence as well, thus taking it even further from the ideal. Thus the core principle of contiguity is still at work in science today, while the increasing difficulties experienced in attempts to show empirical data to be in keeping with it have perhaps led to a reduction in the efforts being made to do so, and in some cases to the disappearance of such efforts altogether. Similar things may be said of Heisenberg's uncertainty principle, its having

---

[68] In keeping with what has been suggested on p. 181 of ibid.

[69] As suggested on p. 61 of ibid.

[70] As is in keeping with the main text, pp. 101–102, and is implied in the previous appendix, p. 245, n. 13 and accompanying text.

received so much attention being a manifestation of the fact that the deterministic form of the uniformity principle lay at the core of science; and as long as there are physicists who try to find ways round Heisenberg's results it may be said still to do so even for quantum mechanics.

Andersen suggests that in the book the question of how principles are modified is only briefly touched upon.[71] Developing this point further then, I should say that what leads to the emendation of particular principles must ultimately be the results of empirical research. The results of measurement indicate the existence of a situation that is theoretically intractable given that the theory or theories in question presuppose the original principle or principles.

How principles are changed can thus be said to be in such a way that the empirical evidence can be explained by a theory or theories based on the emended principle(s).[72] This move may take various forms (such as that of a 'working hypothesis') or take place in certain scientific disciplines but not in others. However, since quantum mechanics constitutes what is presently the fundamental discipline of modern science,[73] if such a change (not merely as a working hypothesis) were to occur there, its influence should penetrate the whole of science – though its effects may only be *manifest* in quantum mechanics itself, since they are too small to appear or be relevant elsewhere.

From this we may go on to note the difference between *emending* a particular principle, such as that of causality coupled with contiguity, and *exchanging* one principle for another, or *removing* a principle altogether. According to my metaphysics of science, it is only in the latter case that one may speak of a change of science (and here it is *not* merely a matter of shelving the principle in question). In this volume I have contrasted major and minor scientific revolutions, major revolutions consisting in a change in one or more principles, and minor revolutions consisting in one theory's being replaced by or developed into another.[74] Here I can clarify this situation further.

---

[71] Andersen (1997), pp. 269–270.
[72] In this regard, cf. *this volume*, pp. 106–107.
[73] In keeping with the hierarchy indicated on pp. 4, 56–57 and 150 of ibid.
[74] Ibid., pp. 104 and 107.

A change of principle which is only an *emendation* of the principle vis-à-vis the core principles of science constitutes a major revolution *within* science, while a change involving an *exchange, addition* or *dropping* of one or more principles of science as a whole constitutes a major revolution *of* science, or what may be termed a *fundamental* revolution. Thus fundamental revolutions (or the having of essentially different principles) separate *sciences*, where major revolutions (or the having of the same fundamental principles interpreted differently) separate *disciplines within* a science. And we may therefore say that, to the extent that there has been an emendation of principles in physics with the acceptance of quantum mechanics, quantum mechanics itself constitutes a discipline different from Newtonian mechanics. But it does not constitute a different *science*.

Note how the notion of *paradigm* is to be applied here. If we accept that quantum mechanics involves an emendation of the principles of physics, though the principles of modern science have not changed (though they have been weakened), those of the discipline of mechanics have changed, and with that change quantum mechanics can be said to have a paradigm different from that of Newtonian mechanics, while the paradigm for science as a whole remains the same. And though the changes in the principles of science's central discipline of physics may have called the paradigm of science as a whole into question, it still functions as the ideal which scientific research generally tries to meet. But even if the paradigm for modern science as a whole should be emended such that e.g. strict determinism were no longer the ideal for scientific research generally, then it would still be against the background of the original deterministic ideal, and in terms of its categories, that the emendation is to be understood, and in that way the original paradigm would still be operative. However, should one or more of the principles of science as a whole be removed or replaced, then we would have a new paradigm for science, and we should no longer say that we still have to do with *modern* science.

We can see the kind of difference that would lead to this sort of change, i.e. to a *fundamental* revolution, if we consider the potential core principles of the various practices or epistemological endeavours treated in Appendix II, where causes can emanate from *spiritual* entities. A change of this sort occurred in the Scientific Revolution, where Aristotle's principle of causality, which allowed teleological and spiritual causes, was *replaced* by the modern-scientific principle of efficient contiguous physical cause.

8. CONCLUSION

As should be evident from the above, all of my reviewers, with the exception of Fleischhacker, have been operating from within an essentially logical-empiricist frame of reference,[75] which to my mind has made it impossible for them to properly appreciate the views being advanced in either of my books. And to this I should add that, quite possibly due to this same predisposition, they have devoted little attention to what I consider to be the books' central aspects.

As regards *Scientific Progress*, no comment is made as to whether or not my theory has succeeded in meeting Kuhn's and Feyerabend's claims while avoiding relativism, including the idea that there can exist *criteria* for judging the relative acceptability of the different aspects of gestalt-switch phenomena, which previously had always been considered to involve a purely arbitrary choice (by Feyerabend, for example). Nor, as regards that book, is there any discussion concerning non-formal conceptual conflict, a notion that is a fundamental aspect of the Perspectivist conception.

As regards the present volume, no mention is made of the central idea that modern science is only one form of science. And the question is not broached as to whether this work succeeds in avoiding the problems met by earlier aprioristic views (such as those of Kant and Whewell), according to which the fundamental principles of science were to be apodeictic rather than hypothetical and constitute the foundation rather than the core of science. Furthermore, science's presupposition of the three particular principles suggested in this book in such a way that its empirical and theoretical aspects are

---

[75] As is in keeping with what I say e.g. on p. 13 of ibid.: "the view [of the Vienna Circle] has been so widely held during the twentieth century as practically to constitute the discipline of the philosophy of science itself; and its influence on those who believe themselves to have moved beyond it is still strong today." Cf. also pp. 43&n., 93 and 185&n.

made clear is nowhere mentioned. Nor, as is related to this, is the question of the importance of distinguishing between the principles of uniformity and causality taken up. Also, the idea that empirical science is empirical in a mensural and not a phenomenal way, and is thus essentially concerned with reality as it is in itself and not reality as we experience it (*pace* Kant), is nowhere discussed. And no one, except Fleischhacker, addresses the pressing social questions raised in Chapter 10.

In sum, my commentators seem unaware of when they are dealing with a philosophical *theory*, whether it be the one they themselves unconsciously adopt (based on the Deductive Model), or the alternatives I have provided in *Scientific Progress* and the present volume. In the case of my own work this is evident from the piecemeal treatment it has received, when what is required is that the views I present each be treated as a *whole*. And the many factual mistakes that have been made in commenting on my work indicate further that it must be read with much greater care. Considering the many problems faced by the theories presently being embraced in the philosophy of science, it seems to me that the views expounded in these two works deserve serious consideration *as comprehensive alternatives* – a consideration they have yet to receive.

REFERENCES

Abernethy, V.
(1993) *Population Politics: The Choices that Shape Our Future*, N.Y.: Plenum Press/Insight Books, 1993.

Agazzi, E.
(1977) 'Subjectivity, Objectivity and Ontological Commitment in the Empirical Sciences,' *Historical and Philosophical Dimensions of Logic, Methodology and Philosophy of Science*, R. E. Butts and J. Hintikka eds., Dordrecht: D. Reidel, 1977.

(1988) 'Science and Metaphysics: Two Kinds of Knowledge,' *Epistemologia* **11**, 1988, Special Issue: 11–28.

(1992) 'Intelligibility, Understanding and Explanation in Science,' pp. 25–46 of Dilworth ed. (1992).

Alcock, J. E.
(1985) 'Parapsychology as a Spiritual Science,' pp. 537–565 of Kurtz ed. (1985).

Alexander, P.
(1963) *Sensationalism and Scientific Explanation*, London: Routledge & Kegan Paul, 1963.

Andersen, H.
(1997), Review of the third edition of *Scientific Progress* and *The Metaphysics of Science*, *Erkenntnis* **47**, 1997: 265–271.

Angel, L. J.
(1975) 'Paleoecology, Paleodemography and Health,' pp. 167–190 of S. Polgar ed., *Population, Ecology and Social Evolution*, The Hague: Mouton, 1975.

Aristotle
The Complete Works of Aristotle, J. Barnes ed., Princeton: Princeton University Press, 1984.

Bacon, F.
(1620) *Novum Organum*, P. Urbach and J. Gibson, trs. & eds., Chicago and La Salle: Open Court, 1994.

Baerreis, D.

(1980) 'North America in the Early Postglacial,' pp. 356–360 of A. Sherratt ed., *The Cambridge Encyclopedia of Archaeology*, N.Y.: Crown Publishers/Cambridge University Press, 1980.

Bailey, C.

(1928) *The Greek Atomists and Epicurus*, Oxford: Oxford University Press, 1928.

Barnowsky, A. D.

(1989) 'The Late Pleistocene Event as a Paradigm for Widespread Mammal Extinction,' pp. 235–254 of S. K. Donovan ed., *Mass Extinctions: Processes and Evidence*, N.Y.: Columbia University Press, 1989.

Bigelow, J. et al.

(1988) 'Forces,' *Philosophy of Science* **55**, 1988: 614–630.

(1992) 'The World as One of a Kind: Natural Necessity and Laws of Nature,' *British Journal for the Philosophy of Science* **43**, 1992: 371–388.

Bird, A.

(1997) Review of *The Metaphysics of Science*, *British Journal for the Philosophy of Science* **48**, 1997: 284–286.

Blackmore, J. T.

(1972) *Ernst Mach: His Work, Life and Influence*, Berkeley: University of California Press, 1972.

(1982) 'What Was Galileo's Epistemology?' *Methodology and Science* **15**, 1982: 57–85.

(1983) 'Philosophy as Part of Internal History of Science,' *Philosophy of the Social Sciences* **13**, 1983: 17–45.

Bok, B. J. et al.

(1975) 'Objections to Astronomy,' pp. 9–17 of Bok & Jerome (1975)

Bok, B. J. and Jerome, L. E.

(1975) *Objections to Astrology*, Buffalo: Prometheus Books, 1975.

Boltzmann, L.

(1899) 'On the Development of the Methods of Theoretical Physics in Recent Times,' pp. 77–100 of his (1905).

(1900, 1902) 'On the Principles of Mechanics, I, II,' pp. 129–152 of his (1905).

(1904) 'On Statistical Mechanics,' pp. 159–172 of his (1905).

(1905) *Theoretical Physics and Philosophical Problems*, Dordrecht: D. Reidel, 1974.

Born, M.

(1951) *Natural Philosophy of Cause and Chance*, Oxford: Clarendon Press, 1951.

Boserup, E.

(1965) *The Conditions of Agricultural Growth*, London: Earthscan, 1993.

Boyd, R.

(1991) 'Realism, Anti-Foundationalism and the Enthusiasm for Natural Kinds,' *Philosophical Studies* **61**, 1991: 127–148.

Braithwaite, R. B.

(1953) *Scientific Explanation*, Cambridge: Cambridge University Press, 1953.

Broad, C. D.

(1949) 'Violations of Basic Limiting Principles,' pp. 37–52 of Flew ed. (1987).

Brush, S. G.

(1968) 'Mach and Atomism,' *Synthese* **18**, 1968: 192–215.

Burnet, J.

(1914) *Greek Philosophy. Part I: Thales to Plato*, London: Macmillan, 1920.

(1920) *Early Greek Philosophy*, London: Adam & Charles Black, 1945.

Burtt, E. A.

(1924) *The Metaphysical Foundations of Modern Physical Science*, London and Henley: Routledge & Kegan Paul, 1949.

Caird, E.

(1877) *A Critical Account of the Philosophy of Kant*, Glasgow: James Maclehose, 1877.

Campbell, N. R.

(1920) *Physics: The Elements*, Cambridge: Cambridge University Press, 1920; reissued as *Foundations of Science*, N.Y.: Dover Publications, 1957.

(1921) *What Is Science?* N.Y.: Dover Publications, 1953.

(1923) *Relativity*, Cambridge: Cambridge University Press, 1923.

(1938) 'Symposium: Measurement and Its Importance for Philosophy – Part I,' pp. 121–142 of the *Proceedings of the Aristotelian Society*, Supplementary Volume XVII, 1938.

(1942) 'Dimensions and the Facts of Measurement,' *Philosophical Magazine* **33**, 7th Series, 1942: 761–771.

Čapek, M.

(1961) *The Philosophical Impact of Contemporary Physics*, Princeton: D. Van Nostrand, 1961.

Caplan, A. L.

(1981) 'Discussion: Back to Class: A Note on the Ontology of Species,' *Philosophy of Science* **48**, 1981: 130–140.

Carnap, R.

(1955) 'Statistical and Inductive Probability,' *The Structure of Scientific Thought*, E. H. Madden ed., Boston: Houghton Mifflin Co., 1960.

Cartwright, N.

(1983) *How the Laws of Physics Lie*, Oxford: Clarendon Press, 1983.

Chang, S. T.

(1976) *The Complete Book of Acupuncture*, Berkeley: Celestial Arts, 1976.

Churchland, P. M. and Hooker, C. A. eds.

(1985) *Images of Science*, Chicago: University of Chicago Press, 1985.

Clark, M. E.

(1989) *Ariadne's Thread*, N.Y.: St. Martin's Press, 1989.

Cohen, M. N.

(1977) *The Food Crisis in Prehistory*, New Haven and London: Yale University Press, 1977.

(1989), *Health and the Rise of Civilization*, New Haven and London: Yale University Press, 1989.

Comte, A.

(1830–42) *Cours de Philosophie Positive*, Paris, 1830–1842, as excerpted in *Auguste Comte and Positivism*, G. Lenzer ed., N.Y.: Harper & Row, 1975.

Crompton, J.

(1992) 'The Unity of Knowledge and Understanding in Science,' pp. 145–159 of Dilworth ed. (1992).

d'Abro, A.

(1939) *The Rise of the New Physics* (2 vols.), N.Y.: Dover Publications, 1952.

Daly, H. E.

(1992) *Steady-State Economics*, 2nd ed., London: Earthscan Publications, 1992.

Darwin, C.

(1859) *The Origin of Species*, Harmondsworth: Penguin Books, 1968.

d'Espagnat, B.

(1992) 'De l'Intelligibilité du Monde Physique,' pp. 111–121 of Dilworth ed. (1992).

Dijksterhuis, E. J.

(1959) *The Mechanization of the World Picture*, C. Dikshoorn tr., Princeton: Princeton University Press, 1986.

Dilworth, C.

(1992) 'The Linguistic Turn: Shortcut or Detour?' *Dialectica* **46**, 1992: 201–214.

(1994) 'Two Perspectives on Sustainable Development,' *Population and Environment* **15**, 1994: 441–467.

(2001) 'Simplicity,' *Epistemologia* **24**, 2001: 173–201.

(2004) 'Il senso comune, i princìpi e la scienza' ('Common Sense, Principles, and Science'), pp. 39–55 of *Valore e limiti del senso comune*, E. Agazzi ed., Milan: FrancoAngeli, 2004.

(2005) 'The Selfish Karyotype – An Analysis of the Biological Basis of Morals,' *Biology Forum* **98**, 2005: 125–154.

(2007) *Scientific Progress,* 4th ed., Dordrecht: Springer, 2007.

Dilworth, C. ed.

(1992) *Idealization IV: Intelligibility in Science*, Poznan Studies in the Philosophy of the Sciences and the Humanities, vol. 26, Amsterdam and Atlanta: Rodopi, 1992.

Divale, W.

(1972) 'Systemic Population Control in the Middle and Upper Paleolithic,' *World Archaeology* **4**, 1972: 222–243.

Duhem, P.

(1905) 'Physics of a Believer,' pp. 273–311 of the Appendix to his (1906).

(1906) *The Aim and Structure of Physical Theory*, P. P. Wiener tr., N.Y.: Atheneum, 1962.

(1908a) *To Save the Phenomena: An Essay on the Idea of Physical Theory from Plato to Galileo*, E. Doland and C. Maschler trs., Chicago and London: University of Chicago Press, 1969.

(1908b) 'The Value of Physical Theory,' pp. 312–335 of the Appendix to his (1906).

Dupré, J.
(1981) 'Natural Kinds and Biological Taxa,' *Philosophical Review* **90**, 1981: 66–90.

Edge, H. L.
(1985) 'The Problem is not Replication,' pp. 53–72 of *The Repeatability Problem in Parapsychology*, B. Shapir and L. Coly eds., N.Y.: Parapsychology Foundation, Inc., 1985.

Einstein, A.
(1948) 'Time, Space, and Gravitation,' pp. 54–58 of his *Out of My Later Years*, N.Y.: Philosophical Library, 1950.

(1949a) 'Remarks Concerning the Essays Brought Together in this Cooperative Volume,' pp. 665–688 of Schilpp ed. (1949).

(1949b) 'Autobiographical Notes,' pp. 3–95 of Schilpp ed. (1949).

Eliasmith, C.
(1998) Review of *The Metaphysics of Science*, *Dialogue* **37**, 1998: 656–658.

Ellis, B.
(1965) 'The Origin and Nature of Newton's Laws of Motion,' pp. 29–68 of *Beyond the Edge of Certainty*, R. G. Colodny ed., N.J.: Prentice Hall, 1965.

(1966) *Basic Concepts of Measurement*, Cambridge: Cambridge University Press, 1966.

(1985) 'What Science Aims to Do,' pp. 48–74 of Churchland & Hooker eds. (1985).

(1992) 'Idealization in Science,' pp. 265–282 of Dilworth ed. (1992).

Ellul, J.
(1964) *The Technological Society*, N.Y.: Random House, 1964.

Enriques, F.
(1906) *Problems of Science*, Chicago: Open Court, 1914.

(1934) *Signification de l'Histoire de la Pensée Scientifique*, Paris: Hermann & Cie, 1934.

306

Evans, D.

(1996) 'Parapsychology: Merits and Limits,' pp. 47–86 of Stoeber & Meynell eds. (1996).

Faye, J.

(1989) *The Reality of the Future*, Odense: Odense University Press, 1989.

Feyerabend, P.

(1975) *Against Method*, London: Verso, 1978.

Field, J. V.

(1987) 'Astrology in Kepler's Cosmology,' pp. 143–170 of P. Curry ed., *Astrology, Science and Society*, Suffolk: Boydell Press, 1987.

Flamm, D.

(1983) 'Ludwig Boltzmann and his Influence on Science,' *Studies in the History and Philosophy of Science* **14**, 1983: 255–278.

Fleischhacker, L.

(1992) 'Mathematical Abstraction, Idealisation and Intelligibility in Science,' pp. 243–263 of Dilworth ed. (1992).

(2002) Review of *The Metaphysics of Science*, *Epistemologia* **25**, 2002: 323–327.

Flew, A.

(1987a) Introduction to Broad (1949), pp. 37–38 of Flew ed. (1987).

(1987b) 'Introduction,' pp. 11–20 of Flew ed. (1987).

Flew, A. ed.

(1987) *Readings in the Philosophical Problems of Parapsychology*, Buffalo: Prometheus Books, 1987.

Frankfort, F. et al.

(1946) *Before Philosophy*, Harmondsworth: Penguin Books, 1949.

Frazer, J.

(1922) *The Golden Bough: A Study in Magic and Religion*, Hertfordshire: Wordsworth Editions, 1993.

Freud, S.

(1916–17) *Introductory Lectures on Psychoanalysis*, London: Penguin Books, 1973.

Fröhlich, F.

(1959) 'V. – Primary Qualities in Physical Explanation,' *Mind* **68**, 1959: 209–217.

Gamow, G.
(1966) *Thirty Years That Shook Physics*, N.Y.: Dover Publications, 1985.

Georgescu-Roegen, N.
(1971) *The Entropy Law and the Economic Process*, Cambridge, Mass.: Harvard University Press, 1971.

Giedymin, J.
(1982) *Science and Convention*, Oxford: Pergamon Press, 1982.

Gingras, Y. and Schweber, S. S.
(1986) 'Constraints on Construction,' *Social Studies of Science* **16**, 1986: 327–383.

Gómez, R. J.
(1992) Review of the second edition of *Scientific Progress*, *Noûs* **XXVI**: 264–270.

Gowdy, J. M.
(1998) 'Biophysical Limits to Industrialization,' pp. 65–82 of M. N. Dobkowski and I. Wallimann eds., *The Coming Age of Scarcity*, Syracuse: Syracuse University Press, 1998.

Graves, J. C.
(1971) *The Conceptual Foundations of Contemporary Relativity Theory*, Cambridge, Mass.: MIT Press, 1971.

Griffin, D. R.
(1996) 'Why Critical Reflection on the Paranormal is So Important – and So Difficult,' pp. 87–117 of Stoeber & Meynell eds. (1996).

Guthrie, W. K. C.
(1965) *A History of Greek Philosophy. Volume II: The Presocratic Tradition from Parmenides to Democritus*, Cambridge: Cambridge University Press, 1965.

Hacking, I.
(1983) *Representing and Intervening*, Cambridge: Cambridge University Press, 1983.

(1984) 'Experimentation and Scientific Realism,' pp. 154–172 Leplin ed. (1984).

Hands, D. W.
(1985) 'Karl Popper and Economic Methodology,' *Economics and Philosophy* **1**, 1985: 83–89.

Hanson, N. R.

(1958) *Patterns of Discovery*, Cambridge: Cambridge University Press, 1958.

Harré, R.

(1967) 'Pierre Simon de Laplace,' pp. 391–393 of vol. 4 of *The Encyclopedia of Philosophy*, P. Edwards ed., N.Y. and London: Macmillan, 1972.

(1970a) 'Constraints and Restraints,' *Metaphilosophy* **I**, 1970: 279–299.

(1970b) *The Principles of Scientific Thinking*, Chicago: University of Chicago Press, 1970.

(1972) *The Philosophies of Science*, Oxford: Oxford University Press, 1972.

Harré, R. and Madden, E. H.

(1975) *Causal Powers*, Oxford: Blackwell, 1975.

Hausman, D. M. ed.

(1984) *The Philosophy of Economics*, Cambridge: Cambridge University Press, 1984.

Heath, T.

(1921) *A History of Greek Mathematics*, N.Y.: Dover Publications, 1981.

Helmholtz, H. v.

(1894) 'Preface by H. von Helmholtz,' pp. xxv-xxxviii of H. Hertz, *The Principles of Mechanics*, N.Y.: Dover Publications, 1956.

Hempel, C. G.

(1958) 'The Theoretician's Dilemma: A Study in the Logic of Theory Construction,' pp. 173–226 of his (1965).

(1962) 'Deductive-Nomological vs. Statistical Explanation,' pp. 98–169 of *Minnesota Studies in the Philosophy of Science* **III**, H. Feigl and G. Maxwell eds., Minneapolis: University of Minnesota Press, 1962.

(1965) *Aspects of Scientific Explanation*, N.Y.: Free Press, 1965.

(1970) 'On the "Standard Conception" of Scientific Theories,' pp. 142–163 of *Minnesota Studies in the Philosophy of Science* **IV**, M. Radner and S. Winokur eds., Minneapolis: University of Minnesota Press, 1970.

(1973) 'The Meaning of Theoretical Terms: A Critique of the Standard Empiricist Construal,' pp. 367–378 of Suppes et al. eds. (1973).

Hempel, C. G. and Oppenheim, P.

(1948) 'Studies in the Logic of Explanation,' pp. 245–290 of Hempel (1965).

Hesse, M. B.
(1961), *Forces and Fields: The Concept of Action at a Distance in the History of Physics*, London: Thomas Nelson and Sons Ltd., 1961.

Hinckfuss, I.
(1998) Review of *The Metaphysics of Science*, *Australasian Journal of Philosophy* **71**, 1998: 130–132.

Hotelling, H.
(1929) 'Stability in Competition,' *Economic Journal* **39**, 1929: 41–57.

Hübner, K.
(1988) 'Metaphysics and the Tree of Knowledge,' *Epistemologia* **11**, 1988: 105–120.

Hull, D. L.
(1981) 'Discussion: Kitts and Kitts and Caplan on Species,' *Philosophy of Science* **48**, 1981: 141–152.

Hutten, E. H.
(1954) 'The Rôle of Models in Physics,' *British Journal for the Philosophy of Science* **4**, 1953–54: 284–301.

(1956) *The Language of Modern Physics*, London: George Allen & Unwin, 1956.

(1962) *The Origins of Science*, London: George Allen & Unwin, 1962.

Hyman, R.
(1985) 'A Critical Historical Overview of Parapsychology,' pp. 3–96 of Kurtz ed. (1985).

Jammer, M.
(1954) *Concepts of Space*, Cambridge, Mass.: Harvard University Press, 1954.

Jerome, L. E.
(1975) 'Astrology: Magic or Science?' pp. 37–62 of Bok & Jerome (1975).

Jevons, W. S.
(1887) *The Principles of Science*, London and N.Y.: Macmillan and Co., 1887.

Johansen, L.
(1979) 'The Bargaining Society and the Inefficiency of Bargaining,' *Kyklos* **32**, 1979: 497–522.

Kahn, C. H.

(1960) 'Anaximander's Fragment: The Universe Governed by Law,' pp. 99–117 of P. D. Mourelatos ed., *The Pre-Socratics*, Princeton: Princeton University Press, 1993.

Kant, I.

(1781, 1787) *Critique of Pure Reason*, N. Kemp Smith tr., London and Basingstoke: Macmillan Press, 1933.

(1783) *Prolegomena To Any Future Metaphysics That Will Be Able To Come Forward As A Science*, P. Carus and J. W. Ellington trs., in I. Kant, *Philosophy of Material Nature*, Indianapolis: Hackett Publishing Company, 1985.

(1786) *Metaphysical Foundations of Natural Science*, J. W. Ellington tr., in I. Kant, *Philosophy of Material Nature*, Indianapolis: Hackett Publishing Company, 1985.

Katzav, J.

(1997) Review of *The Metaphysics of Science*, *Annals of Science* **54**, 1997: 315–316.

Kitcher, P.

(1984) 'Species,' *Philosophy of Science* **51**, 1984: 308–333.

Kitts, D. B. and Kitts, D. J.

(1979) 'Biological Species as Natural Kinds,' *Philosophy of Science* **46**, 1979: 613–622.

Kneale, W.

(1949) *Probability and Induction*, Oxford: Clarendon Press, 1949.

Knight, D.

(1986) *The Age of Science*, Oxford and N.Y.: Blackwell, 1988.

König, G.

(1989) 'Perspektiv,' pp. 363–377 of vol. 7 of J. Ritter and K. Gründer eds., *Historisches Wörterbuch der Philosophie*, Darmstadt and Basel: Wissenschaftliche Buchgesellshaft, 1989.

Krajewski, W.

(1977) *Correspondence Principle and Growth of Science*, Dordrecht: D. Reidel, 1977.

Kripke, S.

(1972) *Naming and Necessity*, Oxford: Basil Blackwell, 1980.

Kuhn, T. S.

(1962), *The Structure of Scientific Revolutions*, Chicago: University of Chicago Press, 1970.

Kurtz, P. ed.

(1985) *A Skeptic's Handbook of Parapsychology*, Buffalo: Prometheus Books, 1985.

Lachmann, L. M.

(1969) 'Methodological Individualism and the Market Economy,' pp. 303–311 of Hausman ed. (1984).

Laudan, L.

(1981) *Science and Hypothesis*, Dordrecht: D. Reidel, 1981.

(1984a) *Science and Values*, Berkeley: University of California Press, 1984.

(1984b) 'A Confutation of Convergent Realism,' pp. 218–249 of Leplin ed. (1984).

Lauener, H.

(1992) 'Transcendental Arguments Pragmatically Relativized: Accepted Norms (Conventions) as an *A Priori* Condition for any Form of Intelligibility,' pp. 47–71 of Dilworth ed. (1992).

Lenski, G. et al.

(1995) *Human Societies*, 7th ed., N.Y.: McGraw-Hill, 1995.

Leplin, J. ed.

(1984) *Scientific Realism*, Berkeley: University of California Press, 1984.

Lindsay, J.

(1971) *Origins of Astrology*, London: Frederick Muller, 1971.

Locke, J.

(1690) *An Essay Concerning Human Understanding*, N.Y.: Dover Publications, 1959.

Lucretius

On the Nature of the Universe, Harmondsworth: Penguin Books, 1951.

Mach, E.

(1883) *The Science of Mechanics*, T. J. McCormack tr., La Salle, Illinois: Open Court, 1960.

(1906) *The Analysis of Sensations*, 5th ed., C. M. Williams tr., Chicago and London: Open Court, 1914.

Machlup, F.

(1955) 'The Problem of Verification in Economics,' pp. 137–157 of his (1978); reprinted from *The Southern Economic Journal* **21**, 1955: 1–21.

(1956) 'On Indirect Verification,' pp. 198–209 of Hausman ed. (1984); reprinted from *The Southern Economic Journal* **22**, 1956: 483–493.

(1960) 'Ideal Types, Reality, and Construction,' pp. 223–265 of his (1978); translated from the original German in *Ordo* **12**, 1960.

(1961) 'Are the Social Sciences Really Inferior?,' pp. 345–367 of his (1978); reprinted from *The Southern Economic Journal* **27**, 1961: 173–184.

(1967) 'Theories of the Firm: Marginalist, Behavioral, Managerial,' *American Economic Review* **57**, 1967: 1–33.

(1978) *Methodology of Economics and Other Social Sciences*, N.Y.: Academic Press, 1978.

Mackie, J. L.

(1976) *Problems from Locke*, Oxford: Clarendon Press, 1976.

Malthus, T. R.

(1798) *An Essay on the Principle of Population*, Harmondsworth: Penguin Books, 1970.

(1830) *A Summary View of the Principle of Population*, pp. 219–272 of his (1798).

Manicas, P. T.

(1987) *A History and Philosophy of the Social Sciences*, Oxford: Blackwell, 1987.

(1992) 'Intelligibility and Idealization: Marx and Weber,' pp. 283–303 of Dilworth ed. (1992).

Martin, P. S.

(1966) 'Africa and Pleistocene Overkill,' *Nature* **212**, 1966: 339–342.

(1967) 'Prehistoric Overkill,' pp. 75–120 of Martin & Wright eds. (1967).

(1973) 'The Discovery of America,' *Science* **179**, 1973: 969–974.

(1984) 'Prehistoric Overkill: the Global Model,' pp. 354–403 of P. S. Martin and R. G. Klein eds., *Quaternary Extinctions*, Tucson: University of Arizona Press, 1984.

Martin, P. S. and Guilday, J. E.

(1967) 'A Bestiary for Pleistocene Biologists,' pp. 1–62 of Martin & Wright eds. (1967).

Martin, P. S. and Wright, H. E. eds.

(1967) *Pleistocene Extinctions: The Search for a Cause*, New Haven and London: Yale University Press, 1967.

Mauss, M.

(1950) *A General Theory of Magic*, London and Boston: Routledge & Kegan Paul, 1972.

Maxwell, J. C.

(1877) *Matter and Motion*, N.Y.: Dover Publications, 1953.

Maxwell, N.

(1984) *From Knowledge to Wisdom*, Oxford: Basil Blackwell, 1984.

McClenon, J.

(1984) *Deviant Science: The Case of Parapsychology*, Philadelphia: University of Pennsylvania Press, 1984.

McCloskey, H. J.

(1971) *John Stuart Mill: A Critical Study*, London: Macmillan, 1971.

McMichael, A. J.

(1993) *Planetary Overload: Global Environmental Change and the Health of the Human Species*, Cambridge: Cambridge University Press, 1993.

McMullin, E.

(1974) 'Empiricism at Sea,' pp. 121–132 of R. S. Cohen and M. W. Wartofsky eds., *A Portrait of Twenty-Five Years: Boston Colloquium for the Philosophy of Science 1960–1985*, Dordrecht: D. Reidel, 1985.

(1984a) 'A Case for Scientific Realism,' pp. 8–40 of Leplin ed. (1984).

(1984b) 'Two Ideals of Explanation in Natural Science,' pp. 205–220 of *Causation and Causal Theories*, P. A. French et al. eds., vol. IX of the Midwest Studies in Philosophy, Minneapolis: University of Minnesota Press, 1984.

(1987) 'Explanatory Success and the Truth of Theory,' pp. 51–73 of N. Rescher ed., *Scientific Inquiry in Philosophical Perspective*, N.Y.: University Press of America, 1987.

Mellor, D. H.

(1977) 'Natural Kinds,' *British Journal for the Philosophy of Science* **28**, 1977: 299–312.

Meynell, H.

(1996) 'On Investigation of the So-Called Paranormal,' pp. 23–45 of Stoeber & Meynell eds. (1996).

314   REFERENCES

Mill, J. S.
   (1881) *A System of Logic Ratiocinative and Inductive*, Books I-III, 8th ed., Toronto: University of Toronto Press, 1973.

Newton, I.
   (1687) *Mathematical Principles of Natural Philosophy*, A. Motte tr., Berkeley: University of California Press, 1934.

Newton-Smith, W. H.
   (1981) *Rationality in Science*, Boston: Routledge & Kegan Paul, 1981.

   (1989) 'The Truth in Realism,' *Dialectica* **43**, 1989: 31–45.

Norris, C. M.
   (2001) *Acupuncture: Treatment of Musculoskeletal Disorders*, Oxford: Butterworth-Heinemann, 2001.

Northrop, F. S. C.
   (1931) *Science and First Principles*, Cambridge: Cambridge University Press, 1931.

Nowak, L.
   (1980) *The Structure of Idealization*, Dordrecht: D. Reidel, 1980.

Oderberg, D. S.
   (1997) Review of the third edition of *Scientific Progress* and *The Metaphysics of Science*, *Ratio* **39**, 1997: 188–194.

Paneth, F. A.
   (1931) 'The Epistemological Status of the Chemical Concept of Element,' *British Journal for the Philosophy of Science* **13**, 1962: 1–14 and 144–160.

Paty, M.
   (1992) 'L'Endoréférence d'une Science Formalisée de la Nature,' pp. 73–110 of Dilworth ed. (1992).

Plato
   The Collected Dialogues of Plato, E. Hamilton and H. Cairns eds., Princeton: Princeton University Press, 1963.

Plutarch
   Plutarch's Moralia, vol. XII, H. Cherniss and W. C. Helmbold trs., London: William Heinemann Ltd., 1957.

Poincaré, H.
   (1902) *Science and Hypothesis*, N.Y.: Dover Publications, 1952.

   (1914) *The Value of Science*, N.Y.: Dover Publications, 1958.

Ponting, C.
(1991) *A Green History of the World*, London: Sinclair-Stevenson Ltd., 1991.

Popper, K. R.
(1962) *Conjectures and Refutations*, N.Y.: Harper & Row, 1963.

Price, G.
(1955) 'Hume's Argument as a Challenge to Parapsychology,' pp. 214–226 of Flew ed. (1987).

Prichard, H. A.
(1909) *Kant's Theory of Knowledge*, Oxford: Clarendon Press, 1909.

Putnam, H.
(1984) 'What Is Realism?,' pp. 140–153 of Leplin ed. (1984).

Quine, W. v. O.
(1951) 'Two Dogmas of Empiricism,' pp. 20–46 of his *From a Logical Point of View*, N.Y.: Harper Torchbooks, 1963.

(1969) 'Natural Kinds,' pp. 114–138 of his *Ontological Relativity and Other Essays*, N.Y. and London: Columbia University Press, 1969.

Reid, T.
(1764) *An Inquiry into the Human Mind on the Principles of Common Sense*, D. R. Brookes ed., Philadelphia: Penn State Press, 1997.

Rhine, J. B.
(1954) 'The Science of Nonphysical Nature,' pp. 23–32 of Flew ed. (1987).

Robbins, L.
(1935) 'The Nature and Significance of Economic Science,' pp. 113–140 of Hausman ed. (1984).

Roberts, N.
(1989) *The Holocene: An Environmental History*, Oxford: Basil Blackwell, 1989.

Robinson, J.
(1962) *Economic Philosophy*, Harmondsworth: Penguin Books, 1976.

Robinson, R. and Broad, C. D.
(1950) 'Two Traditions on the Nature of Man,' pp. 271–276 of Flew ed. (1987).

Ruse, M.
(1976) 'The Scientific Methodology of William Whewell,' *Centaurus* **20**, 1976: 227–257.

(1987) 'Biological Species: Natural Kinds, Individuals, or What?' *British Journal for the Philosophy of Science* **38**, 1987: 225–242.

Salmon, W. C.
(1984) *Scientific Explanation and the Causal Structure of the World*, Princeton: Princeton University Press, 1984.

Samuelson, P. A.
(1972) 'Maximum Principles in Analytical Economics,' *American Economic Review* **62**, 1972: 249–262.

Schilpp, P. A. ed.
(1949) *Albert Einstein: Philosopher-Scientist*, Evanston, Illinois: The Library of Living Philosophers, 1949.

Schmalensee, R.
(1989) 'Inter-Industry Studies of Structure and Performance,' pp. 951–1009 of R. Schmalensee and R. D. Willig eds., *Handbook of Industrial Organization*, vol. II, Amsterdam: Elsevier Science Publishers B.V., 1989.

Sessions, G.
(1995) 'Preface,' pp. ix-xxviii of G. Sessions ed., *Deep Ecology for the 21st Century*, Boston and London: Shambhala, 1995.

Simon, H.
(1979) 'Rational Decision Making in Business Organizations,' *American Economic Review* **69**, 1979: 493–513.

Smith, A.
(1790) *The Theory of Moral Sentiments*, Oxford: Clarendon Press, 1976.

Sober, E.
(1980) 'Evolution, Population Thinking, and Essentialism,' *Philosophy of Science* **47**, 1980: 350–383.

Spector, M.
(1965) 'Models and Theories,' *British Journal for the Philosophy of Science* **16**, 1965–66: 121–141.

Spinoza, B.
(1678) *The Ethics*, pp. 43–271 of B. Spinoza, *Works of Spinoza*, vol. 2, N.Y.: Dover Publications, 1955.

Stearn, W. T.
(1971) 'Linnean Classification, Nomenclature, and Method,' Appendix to W. Blunt, *The Compleat Naturalist*, London: Collins, 1971.

Stoeber, M. and Meynell H. eds.
(1996) *Critical Reflections on the Paranormal*, Albany: State University of New York Press, 1996.

Stokes, D. M.
(1985) 'Parapsychology and its Critics,' pp. 379–423 of Kurtz ed. (1985).

Suppes, P. et al. eds.
(1973) *Logic, Methodology and Philosophy of Science* **IV**, Amsterdam: North-Holland, 1973.

Tait, P. G.
(1876) *Lectures on Some Recent Advances in Physical Science*, London: Macmillan and Co., 1876.

Tambiah, S. J.
(1990) *Magic, Science, Religion, and the Scope of Rationality*, Cambridge: Cambridge University Press, 1990.

Toulmin, S.
(1958) *The Uses of Argument*, Cambridge: Cambridge University Press, 1988.

(1961) *Foresight and Understanding*, London: Hutchinson, 1961.

(1973) 'Rationality and the Changing Aims of Inquiry,' pp. 885–903 of Suppes et al. eds. (1973).

Tylor, E. B.
(1871) *Primitive Culture* (2 vols.), London: John Murray, 1871.

van Brakel, J.
(1986) 'The Chemistry of Substances and the Philosophy of Mass Terms,' *Synthese* **69**, 1986: 291–324.

van Fraassen, B. C.
(1980) *The Scientific Image*, Oxford: Clarendon Press, 1980.

(1985) 'Empiricism in the Philosophy of Science,' pp. 245–308 of Churchland & Hooker eds. (1985).

Veblen, T.
(1909) 'The Limitations of Marginal Utility,' pp. 173–186 of Hausman ed. (1984).

von Weizsäcker, E. et al.
(1997) *Factor Four: Doubling Wealth, Halving Resource Use*, London: Earthscan Publications, 1997.

318   REFERENCES

von Wright, G. H.
(1986) *Vetenskapen och Förnuftet*, Stockholm: Bonniers, 1987.

Wassermann, G. D.
(1981) 'On the Nature of the Theory of Evolution,' *Philosophy of Science* **48**, 1981: 416–437.

Weber, M.
(1903–06) *Roscher and Knies: The Logical Problems of Historical Economics*, G. Oakes tr., N.Y.: Free Press, 1975.

Weinert, F.
(1997) Review of *The Metaphysics of Science*, *Philosophy* **72**, 1997: 330–334.

Weyl, H.
(1949) *Philosophy of Mathematics and Natural Science*, Princeton: Princeton University Press, 1949.

Whewell, W.
(1847) *The Philosophy of the Inductive Sciences*, 2nd ed., London: Frank Cass & Co., 1967.

(1860) *Philosophy of Discovery*, Part 3 (vol. 4) of *The Philosophy of the Inductive Sciences*, 3rd ed., London: Parker and Son, 1858–1860.

Wigner, E. P.
(1967) *Symmetries and Reflections*, Bloomington: Indiana University Press, 1967.

Wilkerson, T. E.
(1988) 'Natural Kinds,' *Philosophy* **63**, 1988: 29–42.

Wilkinson, R. G.
(1973) *Poverty and Progress. An Ecological Perspective on Economic Development*, N.Y.: Praeger, 1973.

Wittgenstein, L.
(1953) *Philosophical Investigations*, Oxford: Basil Blackwell, 1972.

INDEX

Abernethy, V., 220
abstraction (*see also* idealisation), 17,
    45n., 86, 87, *112*, 113, 122, *123*–126,
    143, 145–146
accidents (*see also* properties), 158&n.–
    159, 161, 162
action at a distance, 11, 59, 102n., 205,
    242, 245&n.–247, 249, 251, 257n.–
    258, 260, 262–263
addition, physical, 76, 77
Agazzi, E., 71n., 88n., 119n.
Age of Science, 4, 8, 205
agent(s), 17, 140, 196n., 246, 250,
    258&n., 259&n., 265
agrarian revolution, era, 215, *228–232*,
    236, 238
agriculture, modern, 217, 237–239
Albert the Great, 200
alchemy, 263
Alcock, J. E., 263n., 266&n.
Alexander, P., 105n.
American Medical Association, 268
Amoroso, L., 143
analogue: 24&n.
    abstract, 112
    concrete, 111&n.–113, 143
    source-, 111–113, 143
analogy; analogous, 18, 28, 29, 33,
    105n., 110&n.–113, 143, 144, 198
analytic philosophy, 185n., 264, 277,
    287, 299
Anaxagoras, 55n.
Anaximander, 55n., 203n.
Andersen, H., 275&n., 276&n., 279&n.,
    280&n., 285&n., 296&n.
Angel, L. J., 224n.
animus, 242, 243
a posteriori, 178

apparatus, 46, *77*, 84, 86, 94, 106,
    121n.
applied science, 269
a priori: 60n., 89, 118, 178, 185n., 197n.,
    260n., 261, 268, 286, 287n.
    relative, 71&n., 91, 282&n.–283, 293,
    298
Aquinas, St. T., 200
Archimedes, 202
Aristarchus; Aristarchean, 25n., 200
Aristotle; Aristotelian (*see also* biologi-
    cal philosophy of nature): 3, 52, 54, 55,
    58&n., 59n., 67n., 92n., 93, 99n., 134,
    154, 158&n.–160n., 166n., 198–203,
    206n., 248, 250, 251n., 252, 260n.,
    261–264, 269, 278n., *292*, 298
astrology, 241, *248–251*, 261, 263
atomism, 13–14, 16, 18–19, 21, 24, 54n.,
    55n., 64n., 88–89, 120, 167, 183,
    191, 196n., *197*, *200–203*&n., 208,
    250&n., 260n.–263, 265
Augustine, St., 199

background knowledge, 274
Bacon, F., 56n., 68n., 88n., 201n., 219
Baerreis, D., 223n.
Bailey, C., 54n., 196n.
Balmer, J. J., 106
Barnowski, A. D., 223n.
basic statement, 274
behaviourism, 130, 140n., 144, 187, 264,
    267
belief, degrees of, 174–175
Bell, J. S., 294, 295
Bellarmine, St. R., 25&n., 37, 104n.
Beloff, J., 263n.
Berkeley, G., 16, 27, 185n.
Berzelius, J. J., 130